EXPERIMENTS IN TOPOLOGY

STEPHEN BARR

EXPERIMENTS IN TOPOLOGY

DOVER PUBLICATIONS, INC.
New York

Published in Canada by General Publishing Company, Ltd., 30 Lesmill Road, Don Mills, Toronto, Ontario.
Published in the United Kingdom by Constable and Company, Ltd.

This Dover edition, first published in 1989, is an unabridged and unaltered republication of the work first published by the Thomas Y. Crowell Company, New York, in 1964.

Manufactured in the United States of America
Dover Publications, Inc., 31 East 2nd Street, Mineola, N.Y. 11501

Library of Congress Cataloging in Publication Data

Barr, Stephen.
    Experiments in topology / by Stephen Barr.
       p.   cm.
    Reprint. Originally published: New York : Crowell, 1964.
    Includes index.
    ISBN 0-486-25933-1
    1. Topology.   I. Title.
QA611.B26   1989
514—dc19   88-31661
                                   CIP

DEDICATION

Mathematicians, whose unwonted style
Avoids plain English with the nice excuse:
Readers must learn their language—can beguile
The metaphoric-minded, and induce
Intoxication with ideas as such.
Numbers set indiscretely in a row
Give topologic spaces just as much
As flights of martins in a garden show
Regard for logic. But the martins know
Down is not Up. Topologists ignore
North or South or whether on the floor.
Each has his points; not those who would, instead,
Rather be highfalutin than be read.

Illustrations drawn by Ava Morgan

My thanks are due to Joseph Madachy, editor of *Recreational Mathematics Magazine*, and the editors of *The Scientific American* for permission to reprint some of the puzzles that appear here.

My special thanks to Martin Gardner, who suggested that I write this book, and to Milton Boyd, who taught me the subject.

And to John McClellan, Professor H. S. M. Coxeter, and Professor R. Bing, who gave me aid and comfort.

CONTENTS

EXPERIMENTS IN TOPOLOGY

I

What Is Topology?

Topology is a fairly new branch of mathematics, and it may seem odd to talk of *experiments* in mathematics unless one is, so to speak, at the front line—so advanced that one can hope to make a new contribution—while we are assuming that the reader knows nothing of the subject. But perhaps because it is so new, additions can be made at the side, like branches, if not at the top. Also certain experiments can be made that, while adding nothing, still help one to understand this rather elusive subject.

Topology is curiously hard to define, whereas the following are much less so. *Arithmetic:* "The science of positive real numbers" (*Webster's New Collegiate Dictionary*), or: "The art of dealing with numerical quantities in their numerical relations" (*Encyclopaedia Britannica,* 11th ed.). *Al-*

gebra: "The generalization and extension of arithmetic" (*Enc. Brit.,* 11th). Mark Barr defined mathematics as being "devised to keep facts in abeyance while we dispassionately examine their relations," but this definition applies especially to algebra. *Geometry:* "The study of the [mathematical] properties of space" (*Enc. Brit.,* 11th). Topology started as a kind of geometry, but it has reached into many other mathematical fields. One might almost say it is a state of mind—and is its own goal. (Later we shall see that this last phrase has a topological sound to it.)

In one sense it is the study of continuity: beginning with the continuity of space, or shapes, it generalizes, and then by analogy leads into other kinds of continuity—and space as we usually understand it is left far behind. Really high-bouncing topologists not only avoid anything like pictures of these things, they mistrust them. This is partly because it is not only impossible to make a visually recognizable picture of some of their "spaces," but meaningless. We can, however, get to an understanding of their goal by easy stages, and by looking at certain shapes (or "spaces") from *the topologists' point of view,* if we start with ones that we can see and feel.

A topologist is interested in those properties of a

thing that, while they are in a sense geometrical, are the most permanent—the ones that will survive distortion and stretching.

The roundness of a circle obviously will not: one can tie or glue the ends of a bit of string together and make it into a circle, and, without cutting or disconnecting it, make it into a square. But the fact that *it has no ends* remains unchanged, and if we had strung numbered beads on it they would retain their *order* even if we tied it in knots, provided we count along the string, like a crawling bug (Fig. 1). This would also be true if we used elastic instead of string, because we could only alter the distance between the beads—not their order.

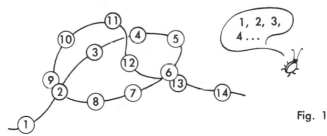

Fig. 1

In projective geometry we get somewhat the same state of affairs: a straight line casts a straight shadow, and a triangle will give a triangular shadow at any angle, even when its own angles change. In topology, though, the straight line doesn't have to remain straight: but it retains the quality of being continuously connected along itself, and with its ends disconnected—or not, as the case may be. (The latter could be so if the line were drawn on a globe, and regarded as straight by the crawling bug, who would report that it did not deviate to either side: like the equator.) It is this connectedness, this continuity, that topology holds on to, and for this reason distortions are only allowed if one does not *disconnect* what was connected (like making a cut or a hole), nor *connect* what was not (like joining the ends of the previously unjoined string, or filling in the hole).

According to this rule, we can take a lump—say round—ot clay and make a cup, but we cannot give it a handle because of the hole in the handle. However we could make both cup and handle from a doughnut-shaped piece (Fig. 2).

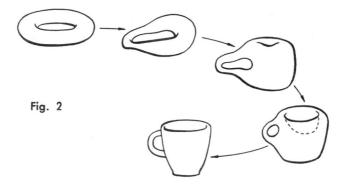

Fig. 2

To be more explicit: we are allowed to make a break, provided we rejoin it afterward. For example, some topologists have said that one can change or distort the first arrangement of a loop of string in Fig. 3 into the second, without alter-

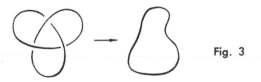

Fig. 3

ing its connectedness. It is true that both are connected the same way, but we obviously cannot do it with string without cutting it and rejoining: but that is allowed. Some say it is possible to do in a 4-dimensional space, but perhaps this modification of the no-cutting-or-joining rule is clearer at this time: Any distortion is allowed provided the end result is connected in the same way as the original.

Another example of this is that one cannot make a flat plate without a hole in it from the doughnut-shaped piece. The latter, incidentally, is called a *torus*. These characteristics—like having or not having a hole—are called *topological invariants*. Sometimes one finds one that turns out to be merely the result of another, but we need not insist on this fact right now.

The lump of clay without a hole is called *simply*

connected, and as a result of being so, we find that, if we draw a circle—or any closed curve—on it (Fig. 4), it divides the whole surface into two: the part inside and the part outside, just as it

Fig. 4

would on paper. The equator does this for the globe, except that it would be hard to say which was the "inside" and which the "outside," but at least it does divide the surface in two.

Now, if we draw another circle, it will either not cut or intersect the first one at all, or it will do so in two places. This means "cut" in the sense of going right through and not merely touching, like the two circles in Fig. 5. This is because if we

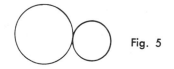

Fig. 5

start drawing the second circle at a point *outside* the first, and then cross over into the *inside,* we cannot get back to the outside to finish the new circle—to join the new line to the point we started at—unless we cross over again. The same applies when we start inside.

Now take the case of the torus (doughnut, Fig. 6). First draw the line *L*. We can see that it has *not* divided the whole surface into two, and so, if we start a second circle at any point, say *P*, this point is neither inside nor outside the circle *L*. Therefore if we cross *L*, the dotted line we are making is not necessarily barred by the line *L* from returning to *P*. As the drawing shows, we *can* have two circles that intersect at *one point only*.

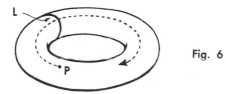

Fig. 6

This fact—not true of a simply connected surface with no holes—is true of anything with a hole, and is a topological invariant.

As was pointed out before, a torus can be distorted into anything with one hole; and a circle into any closed curve that does not join itself anywhere, except for being joined into an endless line. The latter kind are called Jordan curves, after the mathematician who proved that they divide the surface into two distinct regions, which have no

7

points in common but which have the curve as a common boundary—provided the curve is drawn on a simply connected surface: e.g., a plane or a sphere. This may seem to be obvious, but it is unexpectedly difficult to prove. A Jordan curve that divides the surface in two *can* be drawn on a torus, so long as it does not circle the hole, or go through it, as the ones in Fig. 6 do. But on a plane or a sphere *all* Jordan curves divide the surface in two: while on a torus they do not do so necessarily. When one shape or curve can be distorted into another, following our rule, they are said to be *homeomorphic* to one another.

If we draw a triangle on a lump of clay, it is conceivably possible to distort it *homeomorphically* so as to get rid of the three angles and make it into a circle, but if we mark or otherwise identify the apexes as points on the line, they will remain *on it,* and *in the same order* (counting clockwise). Also, if we draw Fig. 7, which is one closed curve joined at two distinct points by another, no distortion that follows our rule can change that description of the

Fig. 7

figure. Not only will the two joints remain as joints, but no new ones will appear, as that would mean making a *new connection*. Thus a sphere with its equator, and another line connecting with the equator at two points, p and p' (Fig. 8), cannot be distorted so that the arrangement of these lines is altered *topologically* (Figs. 9–10).

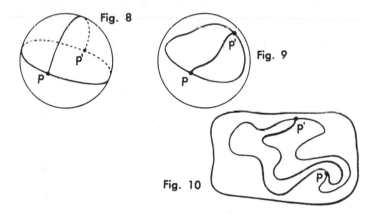

Fig. 8

Fig. 9

Fig. 10

Fig. 9 shows the whole arrangement of lines pulled around onto this side, and bent into arbitrary shapes. (One may distort a drawing on a surface, if we follow the rule.) We see that it still divides the surface into three areas; it still consists of three segments of line, which still meet at two distinct points. These basic facts have survived distortion, while nothing else has. These facts are the kind that topologists are concerned with.

Euler's Theorem

A prime example of topological invariants comes from a theorem the Swiss mathematician Leonhard Euler stated in 1752. It has to do with *polyhedra:* solid geometrical figures, like the cube, or the tetrahedron (Fig. 11), i.e., solids which are bounded by flat planes (*faces*) which have straight *edges* and the edges meet at points, or corners, called *vertices*. You can have more complicated

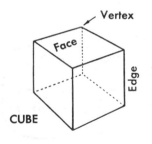
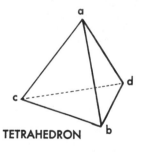

Fig. 11

polyhedra with as many faces as you wish: but never less than 4, as in the tetrahedron. Euler proved that if you add the number of faces to the number of vertices, and subtract the number of edges, you *always* get 2 for an answer. no matter how complex the polyhedron.

Instead of giving the proof, we shall generalize this rule still further in a topological way. The proof will then include Euler's, and be, unexpectedly, easier to follow. First, remembering that in topology we can bend lines, let us draw the tetrahedron on a sphere (Fig. 12). We still have (compare it to Fig. 11) 4 faces (no longer flat but bulging), 6 edges (now curved), and 4 vertices. With

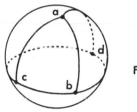

Fig. 12

Euler's rule: 4 faces plus 4 vertices minus 6 edges equals 2. $F-E+V=2$ is the way most books give the equation. Now, as we saw in Fig. 9, page 9, we can pull this whole arrangement of lines around to the front (if we make no breaks or new joints) and get Fig. 13. We still have the 4 vertices, $a, b, c,$ and $d,$ and the 6 edges joining them. Three of the original 4 faces are the triangles 1, 2, and 3,

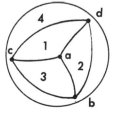

Fig. 13

and the fourth is the space outside the new figure. It is still, topologically speaking, a triangle, as it is bounded by the same 3 edges. This can be drawn on flat paper—all polyhedra can, though in some cases they are hard to recognize—if we remember that the blank space around the figure represents the missing face.

As we said, in topology you can distort if you don't alter the way a figure is connected, and in the case of a polygon, although you may smooth out the angles, you must retain the vertices as *points marked on it*. The pentagon on the left of Fig. 14 becomes the figure on the right, but still has its 5 vertices, and edges: There are certain rules about the way faces, edges, and vertices can be connected in polyhedra—quite complicated—one being that 4 faces is the minimum, another: a vertex is the meeting place of at least 3 edges, and so on, but I am going to generalize Euler's rule to apply to *any* figure we can draw, provided it follows these rules:

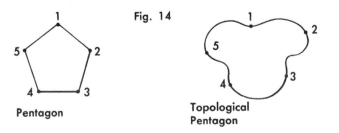

Fig. 14

Pentagon

Topological Pentagon

It must be completely connected: no unattached parts. Every line has a vertex at its free end if there are any free ends, and where it touches or crosses another line—which might be at a previously made vertex. Any enclosure counts as a face, including the outside space. It must be drawn on a simply connected surface—no doughnuts allowed, because then the rule—the formula—changes. We now find that the Euler theorem is, rather surprisingly, easier to prove—or at least to follow, and if we prove the foregoing, we shall have proved it for polyhedra, too. We start with a single line (Fig. 15), and since it has 2 free ends and encloses nothing, it gives 1 face (the space

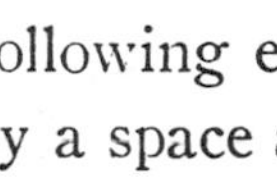

Fig. 15 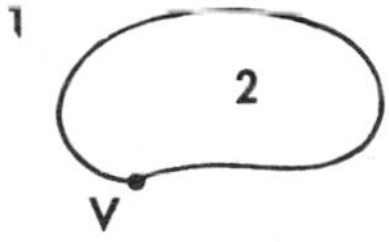   **Fig. 16**

around it), 1 edge (itself), and 2 vertices. In the somewhat unorthodox notation used in the following equations, where a number is followed by a space and then a letter, it indicates what the number is *of;* 2 V means 2 vertices, thus identifying the 2. $1 \text{ F} - 1 \text{ E} + 2 \text{ V} = 2$. If we now join the ends (Fig. 16) it is regarded as making a vertex, which can be put anywhere on the line arbitrarily. This has enclosed a space, giving 2 faces, 1 edge, and 1 vertex ($2 \text{ F} - 1 \text{ E} + 1 \text{ V} = 2$).

13

Now, instead, we cross the first line with another, still enclosing nothing, and we get 1 face, 4 edges, and 5 vertices (1 F—4 E+5 V=2). If they merely *met* we would get 1 face, 3 edges, and 4 vertices: (again getting 1 F—3 E+4 V=2). Also we can put any number of arbitrary vertices on an edge: and each would divide the line into new edges, giving, in Fig. 17, 1 F—4 E+5 V=2.

Fig. 17

When a new line, or edge, meets a loop (a self-connected edge) at its *vertex,* we get 2 F—2 E+2 V=2. If not at the vertex we would have 2 F—3 E+3 V=2. Likewise a line meeting a loop at 2 points gives 3 F—3 E+2 V=2 (Fig. 18).

Fig. 18

It is obvious that the only way to get a new face is by adding at least 1 edge, and this edge must

14

either connect with both its ends or be itself a loop: otherwise it would enclose nothing. Keep in mind that, although in topology we distort things, in the following proof we cannot change anything *after* it has been drawn. The following apply in all cases (or figures).

1. If we add a vertex to an edge between vertices, it divides it: making 1 edge into 2, thus it adds 1 E, canceling the new V, in the expression F— E+V.

Fig. 19

2. Add an edge that meets a vertex—its own vertex on the free end cancels the new edge (in F— E+V).

Fig. 20

3. Add an edge that meets an edge between vertices: it adds 2 E and 2 V (having divided the old edge). These cancel as before.

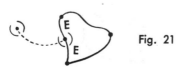

Fig. 21

4. Add an edge with each end meeting a vertex: it adds 1 F and 1 E (but no V) and they cancel.

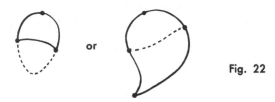

or

Fig. 22

5. Add an edge with both ends meeting the same V: it adds 1 F and 1 E, which cancel.

Fig. 23

6. Add an edge that meets 1 V and 1 E: it adds 1 F, 2 E, and 1 V, which cancel (1 F—2 E+ 1 V=0).

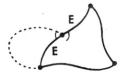

Fig. 24

7. Add an edge that meets 2 edges: it adds 1 F, 3 E, and 2 V, which cancel (1 F—3 E+2 V=0).

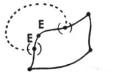

Fig. 25

8. Add an edge with both ends meeting at one V in one edge: it adds 1 F, 2 E, and 1 V, which cancel.

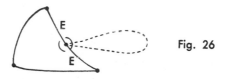

Fig. 26

That exhausts all the ways of adding lines and vertices, and therefore one can draw any figure made of them, and if it is connected, and on a simply connected surface, $F - E + V = 2$. Thus it must be true of polyhedra, also. Try it with a complicated figure drawn at random. We have been stressing the rule that these figures must be on a simply connected surface: what happens to Euler's theorem when they are on a torus? Remembering Fig. 6 (page 7), we can see that it breaks down at once: redrawn in Fig. 27, it shows that 1 F $-$ 2 E $+$ 1 V $=$ 0. And as we said (page 8), a Jordan curve *can* be drawn on the side of the torus and still divide it into two, but not if it circles, or goes

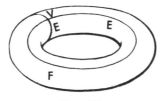

Fig. 27

through the hole. In the same way, any of the connected figures we have just discussed can also be
drawn on a torus and the Euler law applies if they
do not connect either *around* or *through* the hole.
If these lines represent a polyhedron with a hole,
they will do both, and polyhedra were what Euler
had in mind. The simplest polyhedron with a hole
is shown in Fig. 28—made transparent so as to
show all the edges. It has 9 faces, 18 edges, and 9
vertices, giving 9 F—18 E+9 V=0.

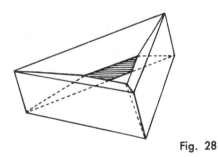

Fig. 28

Without going into the proof, the above is the
new Euler law for doubly connected surfaces, and
it will work for all figures drawn on them provided
we have at least *one line going around the hole,* and
one going through it. NOTE: Euler's law can be
generalized to include any drawing at all that is in
lines and dots: starting with one dot on a sheet of
paper, 1 F—0 E+1 V=2, we can also include

disconnected parts if we change the formula to $F - E + V - n = 2$, where n is the number of disconnected parts (dots, lines, or figures) minus 1. The reader can prove this by experimentation, which will disclose the underlying reason for the formula. The proof turns out to be really quite simple—after we have it. (It applies, of course, only to simply connected surfaces.)

2

New Surfaces

It is important never to be frightened by a higher-mathematician: if he says that something is trivial, he means that it is not general enough —a special case, which is very non-T (or untopological). We must not be put off because he is interested only in the higher abstractions: we have an equal right to be interested in the tangible. He may say that he is for pure mathematics—mathematics for its own sake—not applied—done for the joy of it, while we may derive pleasure from what he would call "mathematical curiosities," like paper models. The tangible is referred to as intuitive, but this is what Paul Alexandroff has to say on the subject (he is a very high-ranking topologist, and we quote from his book *Elementary Concepts of Topology*):

"I would formulate *the* basic problem of set-theoretic topology as follows: *To determine which set-theoretic structures have a connection with the intuitively given material of elementary polyhedral topology and hence deserve to be considered as geometrical figures—even if very general ones."* (His emphasis.)

With this bracing thought in mind, we shall now go to the subject of paper models. These have been objected to on the grounds that paper has a very non-T property: it can't be stretched. This is not absolutely true, but nearly so, and paper is thin enough to serve as a flexible, two-dimensional plane in a model. Its nonstretchability is useful in certain contexts, because it forces us to be rigorous about actual distance, or measurement. Paper also forces us to keep exact track of which side is which, whereas with a lump of clay this might be difficult. We have seen that a surface may be simply connected or not: are there other ways of altering its connectivity? We have said that both a sphere and a sheet of paper are simply connected, but there is nonetheless the difference that the paper is bounded by its edges—as any polygon is—while the sphere is not. Thus although any figure drawn on the sphere can be homeomorphic to one on a plane, the whole surface of the *sphere* is not homeomorphic to the *plane,* because when we stretch the plane over a ball, and bring it around to the other side, we will have a hole which cannot be got rid of without *making a joint*. However, we are now going to consider paper and see how far we can go without the benefit of stretching.

Obviously we cannot make a sphere—although we can make a cube, which is a homeomorphic distortion of a sphere. Also we can make a cylinder: we merely join the ends, or opposite edges AB to $A'B'$ (Fig. 1).

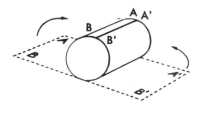

Fig. 1

If our cylinder were longer and flexible we could bring the two ends together, getting a hollow torus (Fig. 2), but by utilizing the limited flexibility

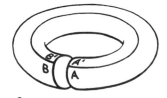

Fig. 2

that paper has, we can make a somewhat distorted one—a true homeomorph of a *torus's surface*. First we agree that a paper cylinder when flattened is a cylinder still: that what we have is connected in the way that a cylinder is, and even if we were prevented from opening it out into its proper round cross section it would be a cylinder from a topological point of view. This being so, we make a long paper cylinder, flatten it and fold it, or bend it so that the ends are facing one another, and then join them with tape (as we did the sides). It is now like a deflated inner tube, flattened (Fig. 3).

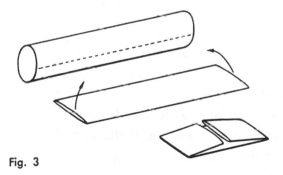

Fig. 3

Notice that the once circular but now flat ends are joined so that the 2 sides—outside and inside—are joined with the inside joined to the inside and the outside to the outside. But if we take a flat strip of paper instead of a cylinder, this unsurprising fact can be frustrated. If we give a half-twist to the strip before we join the ends, we will have the 2 *opposite* sides joined. Also the 2 opposite edges. Fig. 4 is called a Moebius strip, and if we follow along the edge, we will see that it is now in one continuous line which joins itself, forming a loop. And if we make a pencil line on it, running the length, the line will come back to its own beginning.

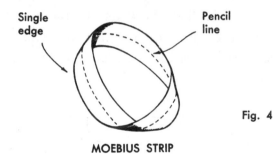

MOEBIUS STRIP

Fig. 4

The Moebius strip has *1 side*—and 1 edge. This is a new kind of surface, and it has a new kind of connectivity. Make one of these, and cut along the center line, and a surprising but perfectly logical thing happens: it stays in one piece even though the cut has gone completely around it and met itself. Logical, because when we gave the half-twist, we joined the upper half of the strip to the lower half (*AO* to *B'O,* and *BO* to *A'O,* in Fig. 5). Our cut has not disturbed this joint. Later we shall

24

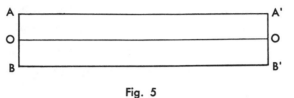

Fig. 5

come back to this, and try some interesting experiments with it. We can say that it has only 1 side, yet the fact remains that we run into semantic trouble when we try to say, "It is impossible to paint the 2 sides different colors because there is only 1 side." *Two* sides? *One* side? . . . which do we mean? This anomaly is by no means trivial and must be cleared up.

Orientability

If we connect the decks of two rafts moored next to one another by pushing them together, we can say that the two surfaces have been joined into one surface, since we could draw a continuous line on the decks from *A* to *B* (Fig. 6). In a similar

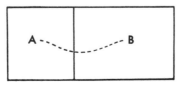

Fig. 6

way the upper side of the strip of paper when twisted and glued has been joined to the under side. But in a different context we can make a dot on it, turn it over and make *another* dot exactly opposite to it. These dots are "on opposite sides" in this new context—although they are on the same side in the sense that we can draw a continuous line from one to the other.

Suppose that the paper were infinitely thin—as a mathematical plane should be—and we have not twisted it. It has an upper side, consisting of an infinite set of points. On the under side is a corresponding set of points—but since the thickness equals zero they coincide with the upper set: *they are the upper set.* Yet we speak of the *2* "sides." If these points individually have no size, how can they have sides? How can they be orientated—right to left, or front to back? Individually they cannot: but in a group they can be—for they are in a particular order, which is reversed when counted, or looked at, from the other *side,* or direction. This is an ex-

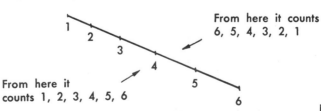

From here it counts
6, 5, 4, 3, 2, 1

From here it
counts 1, 2, 3, 4, 5, 6

Fig. 7

ample of the orientability of a line—a 1-dimensional space (Fig. 7).

Let us now take a strip of paper and cut a hole with a spiral outline. (If we use tracing paper the outline can be drawn very black instead of making a hole.) Provided we do not turn the paper over, the spiral will go clockwise no matter how we turn the paper around. But if we turn it upside down—turn the paper over—it will be reversed in sense, or counterclockwise, like the right-hand side of the figure below.

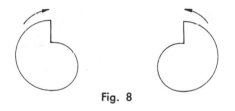

Fig. 8

We could make these holes all over a paper sphere, and seen from the outside they would all be the same: conversely, from the inside they would *all* be reversed. It is easier to show with a paper cylinder, but in both these surfaces we can cover them with similar holes—all clockwise (Fig. 9). If we try this with a Moebius strip (Fig. 10), all goes well for a time—then we find we are next to a hole we made before, and it is counterclockwise, because made from the opposite side, or direction. This means that the Moebius strip is what is called *nonorientable*—which is less open to misinterpretation than saying it is 1-sided. From this we can see that any 2-sided surface is orientable; any 1-sided surface is not.

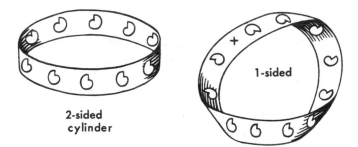

2-sided
cylinder

Fig. 9

1-sided

Moebius Strip
Fig. 10

Dimension

When we say that a solid has 3 dimensions, but a plane only 2, we refer to the fact that mathematically speaking a plane has length and breadth, but no thickness. A plane can, however, be not-flat: e.g., the surface of a sphere. There are two ways of describing this nonflatness: one is to bring in the third dimension of height, and measure the position of all the points of the surface—every part of it. In Fig. 11 the surface is humped in the

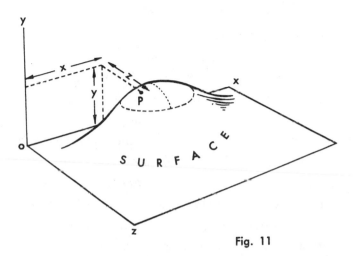

Fig. 11

middle, and we can chart the position of the various points by giving the coordinates, or distances from the point *O* along the three directions *x*, *y*, and *z*. Another version of the 3-dimensional way is to say the surface is that of part of a sphere, or a cone, et cetera.

The other, more topological, method leaves the third dimension out—although it implies it without referring to it. The method is to draw a map of the surface. If we draw a circle on a flat plane and then its diameter, and measure them, we find the ratio of their lengths is 3.14159 . . . (π). If we repeat this on the surface in Fig. 11, the diameter would be too long (see dotted lines). A complete, accurately measured, map of the U.S.A.

that gave every distance and direction could be made as a series of descriptions of where each point is in reference to its neighbors, but it could not be drawn to exact scale on a flat plane: it humps up in the middle—being a sizable section of the globe. It isn't necessary to make points every foot, or even mile: one to a county would show that what was being mapped was not flat. You couldn't put a marker in Iowa and keep on arranging new ones around it so that there was exactly 100 miles between neighbors: Fig. 12 is a hexagonal arrangement, and can be extended indefinitely on a flat plane, but on the U.S.A. as one gets further from the first marker they begin to crowd up: they won't fit any more. Try it on a raw potato with pencil and a tape measure.

Fig. 12

In topology we aim at descriptions that leave out distance entirely, so any surface can be mapped flat, provided it is simply-connected, if we ignore the scale of distances. This is done in atlases—usually using Mercator's projection—but this method at least tries to keep the relative distances as near correct as possible. If we had no way of deducing the world was a ball by direct observation, we could still infer it by measuring points and distances; and if we could not even measure distance, we could at least tell the world was simply-connected by drawing a network all over it and then counting the enclosures (faces), the line segments (edges), and their ends (vertices). If the world were a huge torus we would soon find out there was a hole, because the count would give us $F - E + V = 0$.

So in topology we are concerned with how a surface is connected, and eventually relinquish the idea of distance altogether, but if we do this gradually we can grasp more clearly what it involves.

Two More Surfaces

Up to now we have the following kinds of surface—made from paper—plane, cylinder, torus, and Moebius strip. Parenthetically, the cylinder can be topologically distorted into a plane with one hole. The last shape in Fig. 13 is called an annulus, and it is homeomorphic to any plane with one hole,

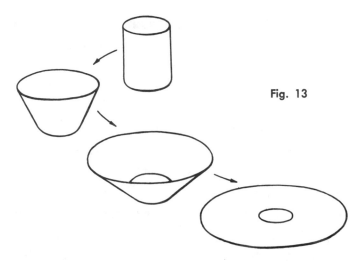

Fig. 13

two edges in all. When we spoke of the paper strip having 2 pairs of edges, we were treating it as a polygon for convenience's sake, but if we want to forget the 4 vertices, or corners, we can in a different context distort it to any closed curve. So let us list the four surfaces mentioned above, according to how they are, or are not joined, and how many sides and edges. They are:

Plane (rectangular):
No joint, 2 sides, 4 edges.

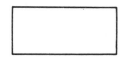

Fig. 14

Cylinder:
  1 pair edges joined, 2 sides, 2 edges.

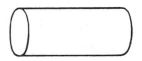

Fig. 15

Torus:
  Both pairs edges joined, 2 sides, 0 edges.

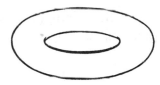

Fig. 16

Moebius strip:
  1 pair edges joined twisted, 1 side, 1 edge.

Fig. 17

Judging purely from the possible combinations of the above operations, there could be two more: both pairs of edges joined with *one* pair twisted, and ditto with *both* pairs twisted. At first sight these seem impossible, but that never deters a true

33

topologist. It is the consideration of all possible combinations of such operations—in fact of almost anything—that is the essence of topology. It is the fact that the combinations can be thought about logically that matters—not whether we can perform them in actuality. As it happens it is possible to make incomplete and imperfect models of the first, and a still more imperfect one of the second. The first is known as the *Klein bottle,* after the German mathematician Felix Klein (1849–1925), and the second is called the *projective plane,* for reasons too heady for this book to go into. Let us take the first first.

The Klein Bottle

What we are asked to do with this is to join the edges AB and $A'B'$—now we label the corners according to which is joined to which. This is what we did for the Moebius strip, but we are now supposed to join the two remaining edges, AB' to $A'B$ (Fig. 18). (Here proportion is ignored, and the

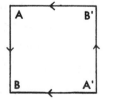

Fig. 18

arrows show the fact that the first joint was done with a twist, and the second one not.) It will be seen, if the reader makes a paper model of this, that it seems completely impossible. Even if we were to add more paper to make this joint, what shape ought it to have? Anything that would fill the bill would have to be shaped somewhat like a man who could take off his coat, turn *one* of the sleeves inside out, put it on again, and button it up around him.

The best way to approach this is to take the joinings in the reverse order: we join the top and bottom first, getting a cylinder. (We imagine a long one.) The arrows are now *directions* around the two circular ends, and when we bent the cylinder and put these ends together—getting a torus—the arrows went in the same rotational direction, or had the *same sense*. Thus we have not lost sight of the lack of twist in the joint: the arrows in Fig. 19 show it, although the fact that all four corners have met at one point conceals it.

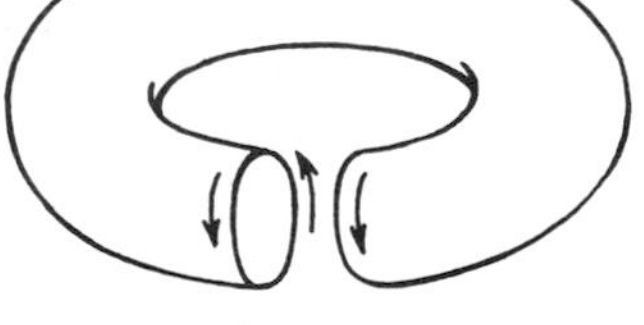

Fig. 19

To make a half-twist in the cylinder *now* would not do what the original half-twist would have done: because the *sense*—say clockwise—would now remain clockwise, even though it would remove A from A'. Thus the direction of the arrows, the sense, is more fundamental, just as the *orientation* of the spiral holes was (page 27). But we have to bring these two end-circles together with their arrows running in opposite senses, and since twisting won't do it now, we must do something else: that is, to put them back to front, which means that one end will have been brought to the other as in Figs. 20–21 (actually from the inside, and the other direction).

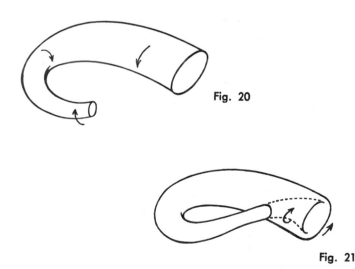

Fig. 20

Fig. 21

As can be seen, the one end has been made narrower and goes through the side. Fig. 21 is the way that the Klein bottle is usually shown: Fig. 22 is another, more symmetrical version. This one

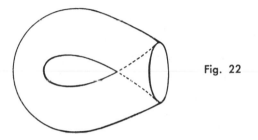

Fig. 22

shows the juncture of the two circular ends as a somewhat acute one, and we must keep in mind that in the ideal case we should not have to resort to such a joint. Still, in the case of a paper model, when two planes meet that way, they can be *imagined* as straightened out (Fig. 23), even when

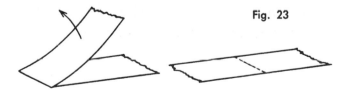

Fig. 23

they cannot, owing to former attachments. This second version has the advantage in being easily constructible with paper, and we shall return to this later. One point to keep in mind: when the two planes are joined as in Fig. 23, the sides facing one another are joined, as are the two sides facing away from one another.

The other new surface, the projective plane, we can leave out for the time being, as the almost fatal imperfection of the quasi-model we *can* make can only be understood in light of further study. For now, we can say that the construction involves not only joining the remaining part of a Moebius strip, but twisting it, also: i.e., *both pairs* of opposite edges are twisted and joined. When we have made some more models of the Moebius strip, with a perfectly arbitrary goal in view, we shall be in a better frame of mind perhaps to understand certain limitations of models, and see beyond them.

Where in the pictures of the Klein bottle (Figs. 21–22) the surface passes through itself, we have to imagine the ends somehow brought together without this intersection—a clear impossibility in real life. The intersection is regarded as being of another kind: the thing happens without making, or needing, a hole. That is to say, in intersecting, *neither plane interrupts the continuity of the other*. In a model this is impossible—in mathematics it is logical, provided we use the right frame of reference. If we think about the surface point by point, we will find no points that are at two places at once, as they would seem to be at the intersection: both on the part of the plane that intersects and the part of the plane that is being intersected, as it were.

This apparent anomaly will become clearer when we get to the question of groups, or *sets*, later on.

3

The Shortest Moebius Strip

Puzzles can sometimes lead to experiments which in turn lead to quite unexpected results. We are now going into the question of actually making a Moebius strip of paper *as short as possible*. By "short" is meant short along one-half its final edge as compared to its width—the width of the edges that we join (after giving the strip a half-twist). What we shall call its length is the length of the flat strip we start with (Fig. 1). The final single edge will be twice this, as it will consist of the upper edge *AB'* and the lower edge *A'B,* joined end to end. The width will be the length of *AB*. Before

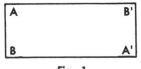

Fig. 1

reading any further, see with an actual paper strip how short it can be and still be twisted—as they say, a half-twist, meaning one end turned over—and then joined.

Without going into a tedious proof we can say that a fold in paper, pressed flat, is a straight line. It also happens to be true that if the paper were *really* unstretchable (inelastic) we could not twist it and retain a perfectly straight line along the middle because in this position (Fig. 2) the edges AB' and $A'B$ are longer than center line CC'.

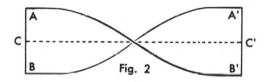

Fig. 2

But we are not going to insist that the strip be held in this position. We can lay the strip flat, and fold it into Fig. 3, which is a triangle, and it can be seen that the strip has been given the required half-twist. We can shrink the triangle until the

Fig. 3

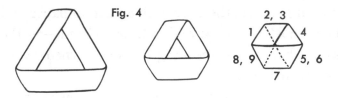

Fig. 4

space in the center is closed (Fig. 4). This would seem to be the limit, but let us make a radical experiment, and start with a strip that is shorter than this proportion—which is easily worked out by geometry, and consists of 9 equilateral triangles joined as in Fig. 5. We take a strip about 2 x 6

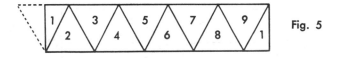

Fig. 5

inches and twist it, and then force the ends together and join them with a piece of cellulose tape. It will look like Fig. 6, more or less, and the place in the center where there had been a triangular opening is now an overlapping.

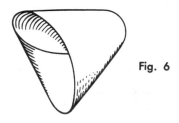

Fig. 6

It will be seen that we can go still further, and shorten it so that it flattens out entirely. When this happens, we have a strip that consists of 3 equilateral triangles, joined as a Moebius strip is. This, it might be objected, is not a rectangle: but we can cut one end off straight, and transfer the piece to the other end, getting a rectangle, and the new-shaped strip is joined as in Fig. 7. This seems like

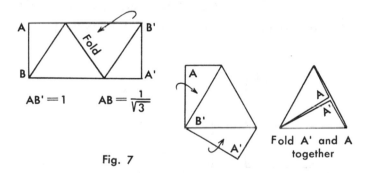

$$AB' = 1 \qquad AB = \frac{1}{\sqrt{3}}$$

Fold A' and A
together

Fig. 7

the limit of shortness: so we shall be still more radical and see what can be done with a square strip. First we notice that we could have made the above folding in a different order (Fig. 8): At stage one

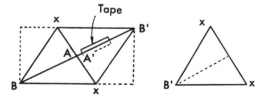

Fig. 8

43

we can fold the small triangles first, join, and *then* fold *xx*. This is identical with the previous one, except that it is not joined—and if it were, the joint would be inside, easy enough to do if we put a piece of tape in the right place before the final fold. Can we do something of this sort with a square?

First we fold it diagonally (Fig. 9), bringing *B'* to *B;* then again on the diagonal *xx*, bringing *A'* up to *A*. We now have the edges *AB* and *A'B'* lined up correctly, and there is just exactly room

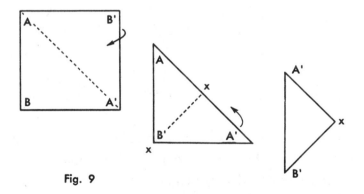

Fig. 9

to join them with tape *across* the intervening edge *AB'*—not really intervening since it does not project beyond *AB* and *A'B'*. What we now have is a Moebius strip that we know is there, even if we cannot, as it were, open it out and look at it. We can, however, cut it down the center—as we did to the strip on page 24—and see what happens. To do this it is advisable to make the cut almost to the ends but not quite (Fig. 10), *before* folding and

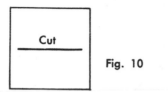

Fig. 10

joining. Then we complete the cuts. Logic tells us we will have one piece, and with a great deal of care—not to tear it—it can just be opened out. It is advisable to use a big piece of paper, and if it does tear, it can be repaired with tape. It opens into a square-folded piece (Fig. 11) that has two sides and is folded over at the four corners in the same way that we would get if we were to flatten the single, two-sided result of cutting the original Moebius strip, page 24. (This arrangement is shown in Fig. 12.)

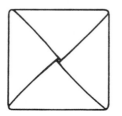

Fig. 11

45

It would have been even easier if we had done it with the triangular strips on page 42. With a folded and twisted square we have not, after all,

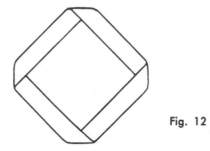

Fig. 12

really followed the method we did at the bottom of page 43. Fig. 13 shows how it would work. First the corners B' and B are brought to the center, then the whole is folded on xx, bringing the sections CB and $C'B'$ in alignment for joining. The remaining sections of the edges to be joined, AC and $A'C'$,

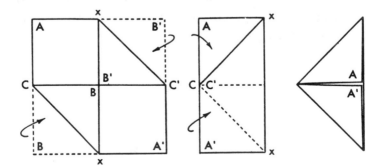

Fig. 13

are now folded together, and joined. What we have
is the same as in Fig. 9, except that the joint is no
longer along the hypotenuse, but doubled along
the center line, half outside, half inside.

This is no improvement, but it suggests a new
attack. Returning to the earlier one at the bottom
of page 43, when we had folded the corners to the
center, suppose we had folded next along the *other*
diagonal, *BB'* (Fig. 14). This would have the ef-

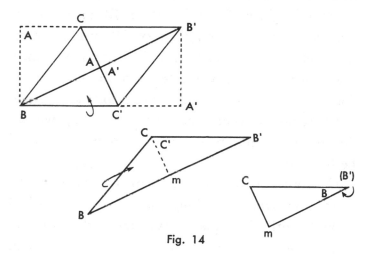

Fig. 14

fect of aligning the sections of edges *AC* and *A'C'*
inside, where they could be joined, and all we have
to do to complete things is to fold along median
Cm, bringing *B* and *B'* together and joining *BC*

to *B′C′*. What we have done here is to make a Moebius strip whose *length,* i.e., distance between the edges we join, is *less than its width*. The actual proportion can be seen to be $w = 1$, $l = 1/\sqrt{3}$ (or 1 by 0.577 . . .)

Can this be improved on? The answer is yes; by an elaborate extension of the method just used, which the reader may care to experiment with, or discover. But there is a way of making an indefinitely wide Moebius strip, recently discovered by Martin Gardner of the *Scientific American* magazine, and given in the appendix. We must say indefinitely wide, rather than infinitely wide because the latter would involve an infinite number of folds, and that would amount to *crumpling,* which is not regarded as cricket. With crumpling we can do almost anything to any shape of paper—it would be like rubber.

Of course none shorter than a square can be opened out when cut along the center—it was bad enough with the square. We know that the resultant 2-sided strip has to have 4 half-twists, and to lie flat when opened out, the available space is filled and the edges meet at the center, Fig. 11, page 45. With a wider strip they would not have room.

The Shortest Moebius Strip

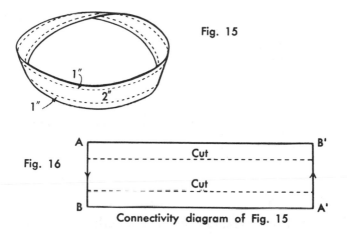

Fig. 15

Fig. 16

Connectivity diagram of Fig. 15

Using the connectivity diagram of a Moebius strip, try this problem: Draw one continuous straight line on the strip so that, when it is cut along, the strip will be in two pieces of equal area. The result of cutting along the dotted line in Fig. 16 will be as described, and possibly surprising, but we make the further proviso that *the cut must start at the edge.*

Answer is in the Appendix.

4

The Conical Moebius Strip

In the last chapter it was taken for granted that we would only consider rectangular strips. If we were to say that the mere connecting of opposite surfaces made a Moebius strip out of any sheet of paper, we could take one with a small projection at one edge, like Fig. 1, turn *A* over and around

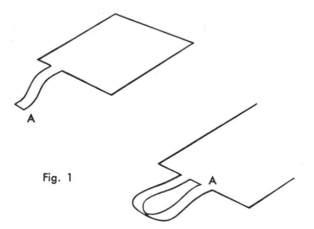

Fig. 1

and stick it to the main part, and call the *whole thing* a Moebius strip. In a sense it would be, but when we get to the more complicated forms, like the Klein bottle or the projective plane, we would be on dangerous ground. For instance, according to directions, all we are supposed to do to make a Klein bottle is to join one pair of opposite edges with a half-twist, and the other pair without twist. Seemingly Fig. 2 would be so connected; and Fig. 3 would be a projective plane.

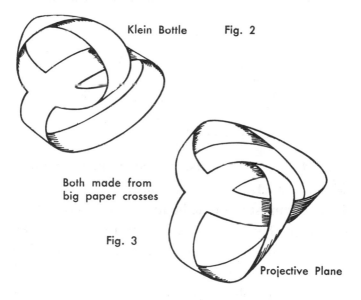

Klein Bottle   Fig. 2

Both made from big paper crosses

Fig. 3

Projective Plane

The trouble here is that in both these cases we are really supposed to connect *all* of both pairs of

edges. Self-intersection has to occur, and we have
sidestepped this issue in Figs. 2–3. (The Klein bottle
and projective plane will be further considered in
the next chapter.)

For now we shall say that some meaningful re-
strictions must be placed on the Moebius strip, too,
as to how much of the edges ought to be joined to
legitimize the transaction. To gain further insight
into this, let us see if the amount of edge involved
can be *increased*. We shall start with an annulus
with a radial slit (Fig. 4). Obviously the ends AO

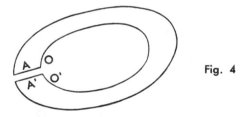

Fig. 4

and $A'O$ can be twisted and joined, making a
Moebius surface out of the annulus. The point is:
How big must the hole be in relation to the outside
diameter? Experiment will seem to suggest that it
cannot be too small, but in fact we don't need *any*
hole to make the right joint. If we take a disk with
a radial cut, fold A over to meet O, and A' *under*
to meet O, we can then join the surfaces on top,
A and B', and also the ones underneath, A' and B

(under *A* Fig. 5). If the creases are rounded out we find a tubular opening running sideways, equivalent to the one in a regular Moebius strip (Fig. 6).

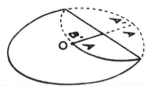

Fig. 5

One edge is -

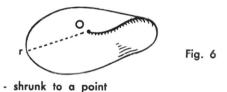

Fig. 6

- shrunk to a point

If this form is cut on a new radius, *rO,* it opens out into Fig. 7. This shape is what we could have started with, and *then* joined the edges *rO* to *rO,* getting the same result, without creases. But do we need to have a full disk (or its two halves, joined)?

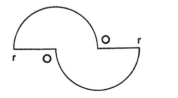

Fig. 7

53

Is it possible to cut an angular segment out instead of the original radial cut, and still make the joint? (See Fig. 8.)

Fig. 8

Translated into the terms of Fig. 7, we would get Fig. 9.

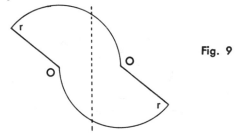

Fig. 9

By folding on the vertical, bring O to O, a fold on Oc would align the edges rO and rO (Fig. 10).

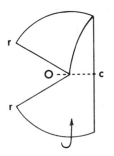

Fig. 10

Experiment will show that these edges can be opened still farther and the joint can be made, but this is not getting us any nearer to *increasing* the amount of edge joined. In fact, the unjoined edge is longer than the joined. So let us return to the first method with the disk with the radial cut, only this time we shall open the cut to a diameter (Fig. 11), and try to make folds to bring the edges (newly labeled) AB and $A'B'$ together: A to A' and B to B'.

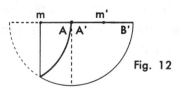

Fig. 11

To do this we fold A to the center (Fig. 12), then the folded part over again, Fig. 13 (turning m up).

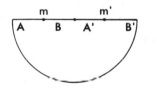

Fig. 12

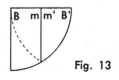

Fig. 13

Finally B' is brought over to B (Fig. 14).

Fig. 14

Perspective

Actually B and A' are contiguous points, but we show them as labels of different corners, or ends of edges, to keep track of which edge is which. Seen from above (Fig. 15), the edges are correctly aligned for joining.

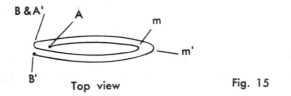

Top view   Fig. 15

The semicircular shape of this has no significance any more—the original annulus is gone—and the joined part is still not as long as the unjoined. Let us see if we can open the space, (originally a slit), to *more* than a straight angle. The arc AB' can be made straight, giving a triangle (Fig. 16). Can this be joined, AB to $A'B'$? And if so, how small can we make α? With the one here

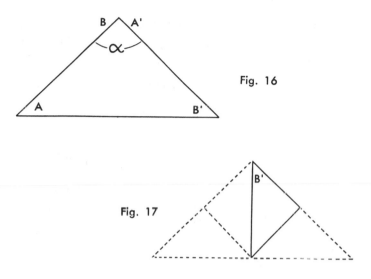

Fig. 16

Fig. 17

($\alpha = 90°$), we can follow the same procedure as above, getting Fig. 17, and the edgewise view will be like Fig. 15. It will be found that the same can be done with an equilateral triangle ($\alpha = 60°$), but when α is made less than that, there is an overlapping of the not-to-be-joined edge AB′, which gets in the way. This, rather irritatingly, is not the limit: α can be reduced to 30° and the Moebius joint can still be made. The procedure is almost as complicated as the last rectangular one—actually more *complex*—and it will be summarily described. Use a tracing-paper model at least 12 inches high, and mark all points on both sides, and color the edges

57

to be joined—on both sides, also. Lay out Fig. 18 in pencil, and cut it out (all points labeled): The apex will be called simply *O,* as the *B* and *A'* of the last model would be in the way. *C* is the midpoint of *AO,* and *E* the midpoint of *BO.*

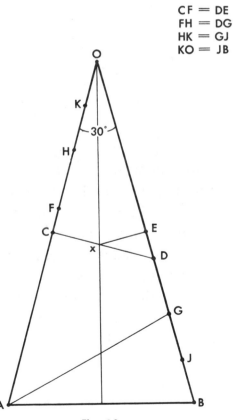

CF = DE
FH = DG
HK = GJ
KO = JB

Fig. 18

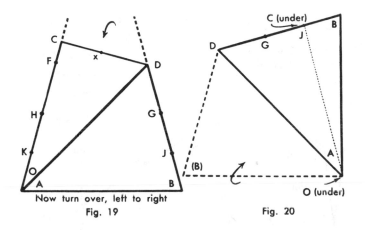

Now turn over, left to right

Fig. 19

Fig. 20

Fold *O* down as in Fig. 19, and turn the model over and fold *BDA* up on *DA* (Fig. 20). *DO* is now behind and flush with *DA*. Fold *BJA* over on *JA* (Fig. 21), then fold *BJA* over on *GA* (Fig. 22). *CA* is now immediately under and flush with *CO* and also flush with section *EO*, to which we join *CA*. Turn model over again.

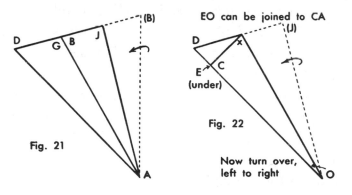

Fig. 21

EO can be joined to CA

Fig. 22

Now turn over, left to right

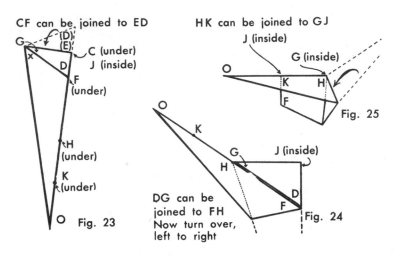

Fig. 23 — CF can be joined to ED

Fig. 25 — HK can be joined to GJ

Fig. 24 — DG can be joined to FH, Now turn over, left to right

Fold *D* down (Fig. 23), and *ED* can now be joined to *CF*, part of section *CO* on the back. We now have section *AF* joined to section *DO* (see Fig. 18), and on the back we have available *FO,* but the corresponding section *BD* is mainly out of sight. We see *DG,* but the rest is flush with *GE,* folded back on itself under it. Nonetheless it can be got at. Fold the lower part up (Fig. 24), bringing *H* to *G,* and join *FH* to *DG.* Turn it all over (Fig. 25), and fold *HKO* to the left, lined up with *GJ* (which is inside). Then fold *KO* back to the right (Fig. 26), and the required edges are in position for joining.

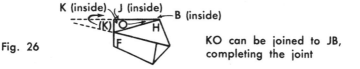

Fig. 26

KO can be joined to JB,
completing the joint

Turned up to show the concealed edge sections (shown diagrammatically in Fig. 27), it can be seen how HK can join to GJ, and, outside of them, KO to JB. This completes the joint. A conical

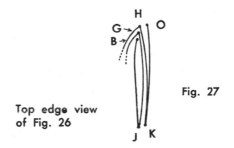

Top edge view
of Fig. 26

Fig. 27

Moebius in name only—but a Moebius nonetheless! Further narrowing of α has the effect of pulling the BJG folds inside, and let us leave them there. Perhaps by the exercise of ingenuity one could still make the joint: I leave it to the reader. The moral of all this is that when we allow *only one kind* of distortion (bending), unexpected relationships persist. Suspicion arises that with *any* distortion allowed, what persists must be invariant indeed, *and perhaps overlooked before.*

5

The Klein Bottle

As was said on page 38, the place where the Klein bottle (Fig. 1) intersects itself is arbitrary, and there are no points on the curve of intersection that are on two different parts of the surface at once, i.e., on both the narrow neck and the main body. The part of the main body where

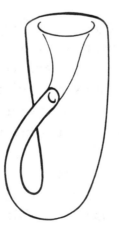

Fig. 1

the hole is, is supposed to continue across it uninterrupted: the hole is there for the convenience of the model.

When followed with the eye, the surface on the outside is seen to connect to the inside of the neck, and thus *all* parts of the surface. Starting at the top rim we can go, bugwise, down on the outside, or down into the neck, and so to the inside. We thus see that the Klein bottle is *nonorientable* (see pages 25–28). We could tell it was, anyway, from its construction—or connectivity—diagram (Fig. 18, page 34), because the half-twist shown by the vertical arrows has joined the front with the back, and no subsequent (or previous) joining—with or without a twist—can alter that fact. The bottle has one side.

But, unlike the Moebius strip, it has no edges: *both* pairs have been joined, and consequently effaced. This is the reason for insisting that the Klein bottle—and, when we come to it, the projective plane—must have *all of both pairs of edges joined,* as stressed before (pages 51–52).

The question naturally arises: What happens when the Klein bottle is cut in two? Does it, like the Moebius strip, remain in one piece, but lose its single-sidedness? Well, the answer depends upon how it is cut, and where. The orthodox model (shown on page 62), when cut down the center symmetrically, yields two odd-shaped pieces that, on examination, prove to be Moebius surfaces: one right-handed and the other left-handed, or mirror images (Fig. 2). The surface on the right is continuously distorted to show its homeomorphic equivalence to a Moebius strip.

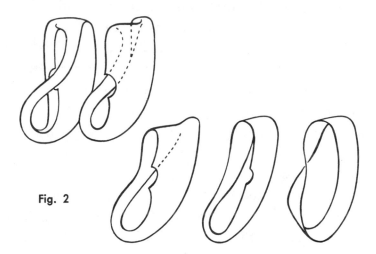

Fig. 2

Actually making this model would be difficult, so let us try paper, and a different version. On page 23 was shown a cylinder of paper, that remains, topologically, a cylinder when flattened. This being the case, we return to the model on page 37, Fig. 22, and see if it can be made *flat* with paper. We make, as before, the horizontal fold

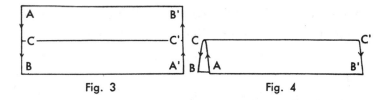

Fig. 3   Fig. 4

(Figs. 3, 4), which would give a cylinder, but no joint is made yet. Then the ends are brought up (Fig. 5) and one put inside the other.

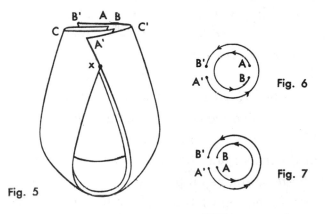

Fig. 5

Fig. 6

Fig. 7

As can be seen, the corners $A'B'$ are not in contact with AB, but as pointed out (pages 35–36), this is academic in the case of the cylinder, and, for the same reasons, also in this case, a Klein bottle. The *sense* (rotational) of $A'-B'$ is correctly aligned with $A-B$. What we now have, looking down on the ends, is shown diagrammatically (Fig. 6), and it is the same as Fig. 7 where the corners $A'B'$ are in contact with *AB,* from the point of view of *sense.* Once an edge is distorted (and joined) into a circle, the only criterion is its sense or orientation.

In Fig. 5 we join the edges $A'C'$ to CB, and $B'C'$ to CA, leaving an opening between CB and CA the opening at the top of the Klein bottle. The self-intersection is from x to C. The two previously unjoined sections of the long edges (parts of AB' and BA' below x, and also inside above it to C') can now be joined, and we have a paper model of a *symmetrical* Klein bottle.

If we do not make this final joint—of the long edges—but regard the lack of joint as part of a new cut which we continue along the longitudinal fold CC', we get the two Moebius strips mentioned earlier: one the reverse image of the other. We note again, as on page 37, that the acute-angle joints at the top of these Moebius strips (Fig. 8)

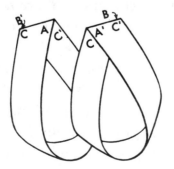

Fig. 8

—and the Klein bottle above—can be (mentally in the case of the latter) opened out into a smooth curve.

Fig. 9 shows the homeomorphic distortion of this model into the more usual depiction of a Klein bottle. On the last drawing a dotted line shows where the cut should be to get the two Moebius surfaces.

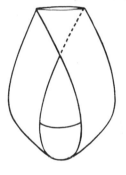
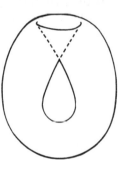

Fig. 9

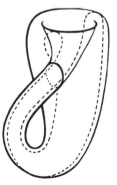

Keeping in mind the principle of correctly orientated joining of the edges AB and $A'B'$, let us put
them together in a different way: in fact according to Fig. 7. To do this means giving a half-twist
to the $A'C'B'$ part (Fig. 10). This, we know, does
not alter the sense of the orientation, but if we
now, after joining ACB and $A'C'B'$ as before,
merely refrain from joining the edges AB' and
$A'B$ at all, and consider this *alone* as our cut, and
open it out, we have a *single Moebius strip*—folded
longitudinally, and with an acute transverse fold
at $AA'–BB'$.

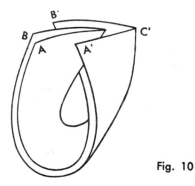

Fig. 10

Translated to the usual model, the cut is tricky
to find: it is shown in Fig. 11, and we can see that
it does *not* divide the Klein bottle into separate
pieces. We show it successively being opened out
in Figs. 12–14.

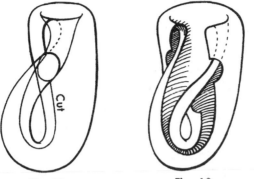

Fig. 11   Fig. 12

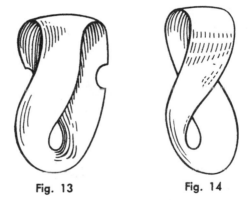

Fig. 13   Fig. 14

Two more experiments can be performed with this method. The cuts have been made along the natural, or obvious, lines of existing folds or edges. What happens when we make them diagonally or, more accurately, spirally? Let us take the first paper model we made—which, when cut, gave two Moebius strips—only this time run the cut from the right-hand corner, $AC'B$, down, around, and up to x, instead of to C as we did before. This cut will have to be on the surface facing us, $A'C'-CB$, *only;* otherwise nothing very interesting happens (try it and see). Then on the back surface, $AC-C'B'$, we make another cut from C, down around and up to x. Bear in mind that this is being done to a model that has had *all* the necessary joints made, and that, in making the second of these new cuts, when we get to x we are at the intersection, and the joint ceases along the portion $xC;$ but with the first cut, when we get to x, we must continue up inside to B (Fig. 15).

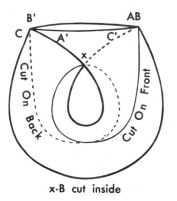

x-B cut inside

Fig. 15

What we have done is to make *two* separate cuts: each self-connecting like a loop, but not really connected with one another from the point of view

of a true Klein bottle, as they merely pass one another at the intersection, and are on *different parts*
of the surface as a whole. The result is quite unexpected: the two cuts have left the Klein bottle
in one piece, which we will find has *2 sides* and,
therefore, *2 edges*. It will be seen that this really
amounts to cutting as on pages 68–69, getting a
Moebius strip, and then cutting *this* lengthwise,
so the resultant single loop is to be expected.

Fig. 15 has been distorted at the lower part, as
though the paper were rubber, and the turning-
around-and-up of the folded strip is flattened out,
without perspective. The two cuts make the 2 edges
of the resultant single piece. Fig. 16 shows the
translation of these cuts to the usual model. It will
be seen that one of the cuts goes *through* the inter-

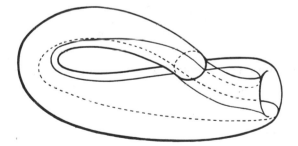

Fig. 16

section, in theory, so here it goes around the neck, following the arbitrary closed curve of the hole, on both sides. (The neck is not joined to the edges of the hole.) If the reader is interested in making a nonflat model along the lines of the above, short of blowing it out of glass (which would be practically impossible to make cuts in), the best thing is a paper model like Fig. 17 which, although it resembles a steam cabinet more than a smooth Klein bottle, is topologically the same as the latter.

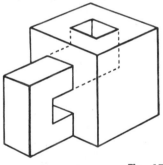

Fig. 17

It also has the advantage of being cuttable. (A steam cabinet, one imagines, for the man mentioned on page 35, who could put on his coat with the sleeve turned inside out.) A scale pattern for its construction is given in the Appendix.

The second of the remaining experiments is to make the rim connection, $AC–C'B'$ and $A'C'–CB$, with the edges alternating. The top view shown in Fig. 18 gives the method and it is extremely hard to visualize this in terms of the usual model. The

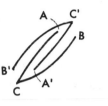

Fig. 18

part we have been calling the neck is no longer all inside the main part: half of each part is inside and half of each is outside. With an ideally imagined Klein bottle, in one respect it makes no difference to the way in which the *inside* is joined to the *outside*. As shown in Fig. 19, the two planes are each marked *A* and *B,* indicating the two sides to be joined. When the joint is as on the left, *A* joins to *A,* and *B* to *B*. Fig. 20 shows that the mere

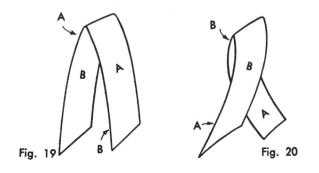

Fig. 19   Fig. 20

reversal of *position* does not affect this: *A* still joins *A,* and *B* still joins *B*. However, when we make the model this way, and cut it symmetrically down the edges or folds, as with the first model we made (page 66), instead of getting two Moebius strips that are mirror images of one another we get two that are *identical*. On the other hand, when cut diagonally, or spirally, as on pages 69–71, the result is *one* strip, with 2 sides, 2 edges, and *2 half-twists*. (The reason for this will be seen later.)

To represent cuts made on this surface in terms of the orthodox model is meaningless, as the latter no longer gives a true picture of the *new manner of intersection*. The nearest equivalent along the lines of the smooth one is shown in Fig. 21: it has, like the others, only one side, but the joining and

Fig. 21

intersection follow very arbitrary rules. It really does not belong in polite topological society and should only be used to annoy mathematicians who disapprove of paper models. (It is called the Slipped-disk Klein bottle.)

Let us enumerate the various models we have made, and their resultant pieces when cut. We shall do this in the form of the connectivity diagram we have used before. Bear in mind that the different arrangements of the joints, and the angles of cutting in these models, has, with the exception of the slipped-disk one, not affected the Klein bottle as such—only the way in which we have cut it. Diagrams of each model and cut are shown.

1. *Pages 63–65*: 1 cut along *AB′* and *CC′*. Gives 2 M-strips, mirror images (Fig. 22).

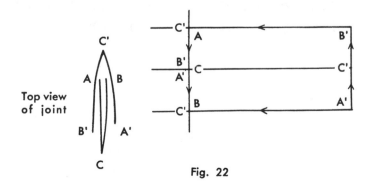

Fig. 22

75

2. *Page 68:* 1 cut (failing to join *AB′* to *A′B*). Gives 1 M-strip (Fig. 23; see note later).

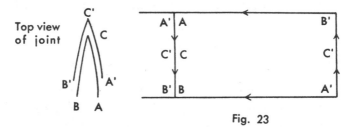

Fig. 23

3. *Pages 69–71:* 2 cuts, *Cx′* and *xC′*. Gives 1 strip: 2 sides, 2 edges, 4 half-twists (Fig. 24).

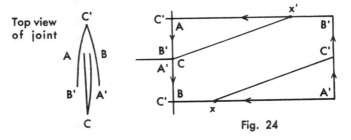

Fig. 24

4. *Page 70:* Not shown. Cut is through both surfaces, gives a Polynesian canoe (Fig. 25).

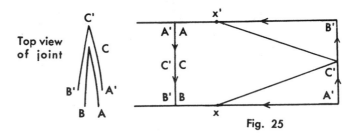

Fig. 25

5. *Pages 73–74:* I cut along *AB'*, but as *AC* has been joined to *C'B'* with sense opposite to joint of *CB–AC'*, it gives 2 M-strips with the *same* orientation, i.e., identical (Fig. 26).

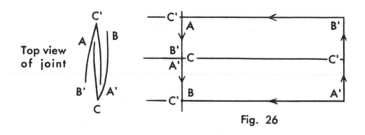

Fig. 26

6. *Page 74:* 2 cuts, *CB'* and *xC'*. Gives 1 strip: 2 sides, 2 edges, 2 half-twists (Fig. 27). When opened out and flattened, gives the folds as shown in Fig. 28.

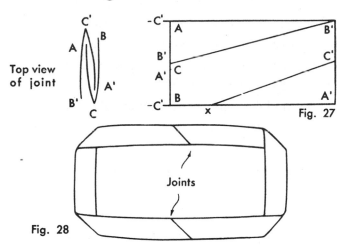

Fig. 27

Fig. 28

77

6

The Projective Plane

In the Klein bottle we had to get around the difficulty of an intersection by cheating a little in the proportions, but in the projective plane the intersection is total, unavoidable, and involves all of both pairs of edges. This is because *both* pairs are given a half-twist when joined. A paper model —or sort of a model—can be made with a considerable amount of trouble, and when it is made it cannot be opened up, like the square Moebius strip. All one can do is say, "My!"

The model in question is made the same way as the square M-strip, except that we not only join AB to $A'B'$ across the intervening edge, as on page 44, but the rest of the edge (in two folds) is joined, as it were, *through* the first joint. To do this we have to resort to the trick shown in Fig. 1, a small sample of the fake intersection being given at the

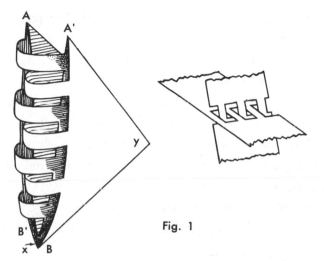

Fig. 1

side. There seems to be a question among topologists as to the legitimacy of this model: the point *x* at which the intersection *intersects itself* is a little dubious, and might cause raised eyebrows in strict circles. If a projective plane has a hole in it—even if the hole is merely the removal of one point—it is deformable into a Moebius strip, even as a sphere is deformable into a plane if a hole is made. This deformation, however, is not an ordinary one: it would involve diametrical lines composing the whole surface passing through the others. When the hole is made (but we have not begun to distort), the surface is what is known as a cross-cap.

In a sense one could make a cross-cap merely

by cutting the corner *y* off the model on page 79, but a better way to visualize it is to imagine—or make, but it would be quite hard—a Moebius strip in which the edge consists of fairly heavy wire, and the surface is replaced by strands of very elastic rubber—say thin rubber bands cut open— which are attached crosswise. Fig. 2 shows the arrangement. If the wire edge is labeled with numbers in pairs on the points of the edge that are opposite one another, and we open the wire in the way shown in Fig. 3, the strands are stretched, but

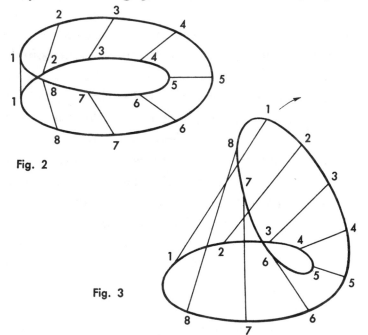

Fig. 2

Fig. 3

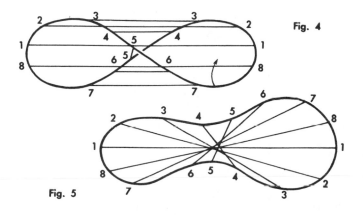

Fig. 4

Fig. 5

as can be seen (Figs. 4–5), they will have to pass through the wire and one another as we approach a circular form. If we imagine they have done this, when a circle of the wire is made, every strand is now a diameter. Consequently the strands are all stretched the same amount, and if we imagine they were truly continuous and *joined sideways,* as they would be if they formed the original surface, we would now have a wire-edged rubber disk (Fig. 6). It would have no hole to correspond to the loop, 1 edge as before, and *1 side.* This is because if the orientation of the new edge is examined, we see that the numbers run in such an order that opposite parts of the edge are connected with a half-twist, which means *the 2 sides are connected.* Now we distort the disk into a square (Fig. 7), and the

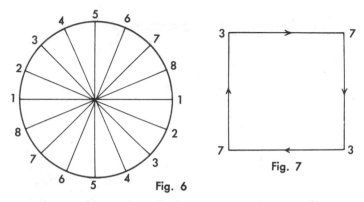

Fig. 6

Fig. 7

numbers show that it is connected according to the recipe for joining the edges of a projective plane. All we have to do now is to let the square into the side of a sphere, like a patch on a tire, and we have a somewhat distorted projective plane: really a sphere with one cross-cap.

The significance of all this will seem a bit remote, but we shall next offer an example of the use of models which shows them as a means of analyzing forms, as well as merely exhibiting them.

Symmetry

From now on we shall call the "half-twist" given to a Moebius strip just "a twist"—the strip has been turned over *once,* and "twist" is clear enough, and saves words.

Some of the questions raised by models are not only hard to answer in set-theoretic topology, but might not even arise. For example, why does a Moebius strip with only 1 twist give, when cut down the middle on its axis, a loop with 4 twists? Also it seems, to say the least, peculiar that the cylinder and torus are symmetrical, while the Moebius strip is not, since it can be either left- or right-handed. Cylinders and tori can be made asymmetrical, but the point is that they *can* be symmetrical, whereas a Moebius strip—or at least its model—cannot be. While a Klein bottle can be cut so as to yield a single Moebius strip, we can get either a left-handed or a right-handed one, depending on which side of the neck, or intersection, we cut. How about a projective plane?

Martin Gardner sent this writer a small paper model, explaining it was a projective plane. It looked simple, easy to make, and as innocent as a mushroom. It is shown here, Figs. 8–11, and this, and all subsequent models should be made by the reader.

The square of paper (Fig. 8) has two cuts, xx'–ee' and yy'. In Fig. 9, the right half BA' is folded to the left on the dotted line; B on A, and A' under B'. This is done by tucking the slit yy' into xx'. In Fig. 10, the bottom flap B' is folded up onto

B, and in Fig. 11, the flap A' is folded up behind A. Now the top and left-hand edges of the flaps A and A' are joined, and so are the top and left-hand edges of flaps B and B', and the edges of the vertical portion of the cut ee' are rejoined. The unjoined edges xx' and yy' are considered as intersecting the uncut surfaces z and z', and could in fact be partially joined in the way shown in Fig. 1 (page 79).

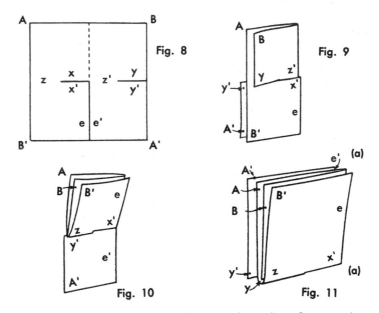

Fig. 8

Fig. 9

Fig. 10

Fig. 11

Examination shows that each pair of opposite edges of the square, Fig. 8, has been joined with a twist, and the slit *ee'* is reconnected. Thus, if we allow an imperfect intersection, as we did with the Klein bottle models, we have a projective plane. All seems well, but the edge *BA'* is half in front and half behind the edge *AB'*: is this allowable? Let us try an experiment.

We shall make only the joints for the top and bottom edges *AB* and *A'B'*. Things will be clearer if we use a very narrow rectangle, shown on its side, and cut and labeled the same; Fig. 12. The

A' ┌──────────────────────────────────────┐ B
   │ ┌──────── e ────────┐   │
   │ └──────── e' ───────┘   │
B' └──────────────────────────────────────┘ A

Fig. 12

result should be a Moebius strip when *AB* and *A'B'* are joined, but when we try to open it out it has a most unorthodox appearance—we have, of course, not forgotten to re-join *ee'*—Fig. 13. Perhaps it is a Moebius strip in disguise—the homeomorph of one?

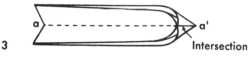

Fig. 13   Intersection

We now cut it along its center axis *aa'*, the elongated equivalent of *aa'* in Fig. 11, and we should have a single 2-sided, 2-edged loop with 4 twists. Instead we have Fig. 14, and when pulled out sideways—disengaging the intersection—we find only 2 twists. Something is wrong.

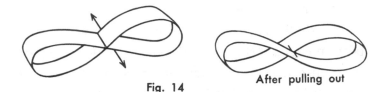

Fig. 14   After pulling out

We repeat the experiment, joining only the other pair of edges, *AB'* and *BA'*, and this time using a rectangle narrow in the other direction, Fig. 15. Having also re-joined *ee'*, we cut along *aa'*, and

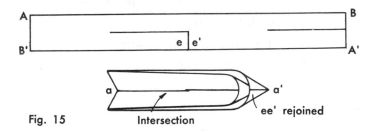

Fig. 15   Intersection   ee' rejoined

we have a loop with *no* twists: a cylinder. What has happened? Well, for one thing we have eliminated the intersection by cutting through it, but why should this cancel the twist?

Let us return to Gardner's square version,

shown in Figs. 16–19. When we first tuck the slit
yy' into *xx'* (Fig. 16), we shall refrain from creas-
ing the paper. Next we bring *A'* up behind *A*, and
B' up in front of *B* (do not join *ee'*), and seen
from above (Fig. 17), it is a double cone. In Fig.

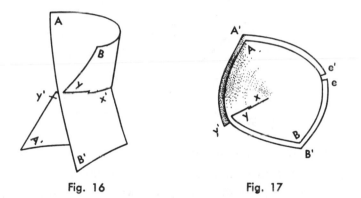

Fig. 16   Fig. 17

18 we have cut off the corners *AA'* and *BB'*, getting a smooth-edged cone. It is homeomorphically
the same as Fig. 11. We open it out again and have
a disk, Fig. 19. This suggests a change that can

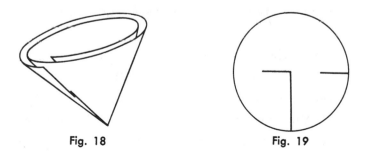

Fig. 18   Fig. 19

be made in the original square model, shown in Fig. 20: the dotted lines are the old edge positions.

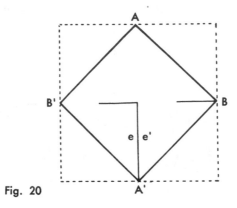

Fig. 20

When folded as before, there is a new line-up of edges, in that now there is no edge half behind and half in front of another edge. When we make the same tests as we did above, by joining only one pair of opposite edges at a time, we get a slightly different result. We will have to use square models for this, as the elongated versions do us no good with the new diagonal slits. This time we get a loop with 2 twists in both cases—no cylinder this time.

It is rather hard to see what we are doing with the square models, but with a pencil we can trace through them and count the number of twists. Since this time we did not cut along the intersection, it is not so surprising that we did not get a cylinder as we did before, but why *2* twists, instead of 4? Let us try again.

In Fig. 18 we erased the distinction between the pairs of edges joined, by cutting off the corners, making the edges continuous. The ideal projective plane has no corners, and neither does a cross-cap —the one produced by opening out a Moebius strip, pages 81–82, has none, although if it did it would still be a cross-cap. The ideal versions of all these forms are completely symmetrical—even the Moebius strip, odd as it seems. Let us take the smooth-edged cone, Fig. 18, and flatten it out by adding more paper. We cut down the dotted line, Fig. 21, through both layers, open out and add gussets, which are V-shaped (Fig. 22). If these are wide

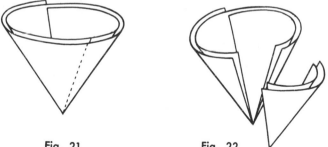

Fig. 21   Fig. 22

enough the cone will become a disk—without topologically altering it. We shall have *2* thicknesses running in a flat helix, and with an intersection

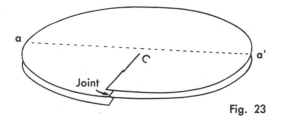

Fig. 23

running radially from center to edge, Fig. 23. It is easier to make this model if we start from scratch with 2 disks, cut radii in each, and join. To make this into a projective plane we join the circular edge of the upper to that of the under surface. If we now cut along the line *aa'*, just beyond the center *C*—the end of the intersection—we get 2 pieces, shown in Fig. 24, with outer edges already joined,

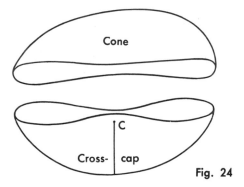

Fig. 24

and tilted for a better view. The bottom half is a cross-cap and the top is topologically a disk. As we said, page 82, fitting a cross-cap into a hole in a sphere gives a projective plane, and our disk is topologically a sphere with a hole in it: so before cutting *aa'* we had a projective plane.

The difference here is that the cross-cap on pages 81–82 was flat, of one thickness, and had its intersection concentrated at the center—and we could not make it in actuality. It was symmetrical about a point in a plane, but the fact that the present one is not is no worse than the models we made of Klein bottles—we must do the best we can with intersections. Both they and the model we have just made are symmetrical about a line drawn on them, or a plane in 3 dimensions, but a paper Moebius strip is not. Let us digress to a case where intersections are not merely inaccurately represented, but side-stepped entirely.

On page 51, Figs. 2 and 3, we showed very incomplete models of the Klein bottle and the projective plane made from paper crosses: incomplete because they are supposed to have *all* of both pairs of edges joined. In a similar way we could join a square Moebius strip by cutting slits in the sides, Fig. 25. It will twist easily about the tiny uncut part, *P,* but the slits are part of its outline, so it is not fair and square. Is the slit *ee'* in the Gardner models akin to this? Hardly; since it gets re-joined, whereas the slits in Fig. 25 cannot be, after join-

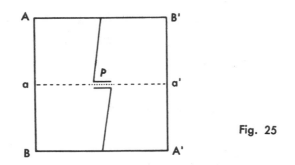

Fig. 25

ing *AB* to *A'B'*. In spite of this defect, the latter will give a loop with 4 twists when cut along its axis, *aa'*.

We shall now make axial cuts in the cruciform model of page 51, shown here again, but flattened to clarify the twists, which become folds, Figs. 26–27. These are both of projective planes, Fig. 26 being symmetrical in that one pair of arms are joined with left-hand twist and the other pair with a right-hand twist. In Fig. 27 they are twisted the same way: left. After cutting on the dotted lines, Fig. 26 gives 2 loops, linked. One of these has 2 left-hand twists, and the other 2 right-hand twists. Notice: 2, not 4.

When Fig. 27 is axially cut it gives (shown next to it) 2 unlinked loops, one with 4 twists, and one with none—a cylinder. This series of experiments is reminiscent of the Gardner model, and suggests that *it* is symmetrical. When we trim the

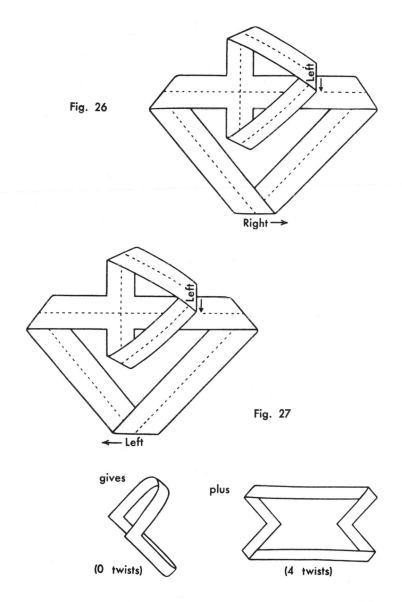

Fig. 26

Left

Right →

Left

Fig. 27

← Left

gives

plus

(0 twists)

(4 twists)

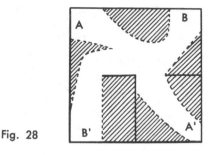

Remove
shaded
areas

Fig. 28

latter to the bone, as it were, Fig. 28, making it
cruciform but distorted, it is plain that the two
pairs of edges are joined differently: right and
left. (In trimming it is wise to make the creases
for the folds first.) The reader may want to try
the experiment—after reading this chapter—of
joining adjacent, rather than opposite, arms of
cruciform models, with various arrangements of
twists and cutting. Some of the results are quite
bizarre. We leave it to the reader to decide what
they are.

We have still not explained why some of these
models give only 2 twists when cut, but for that
matter, why does a Moebius strip give 4? It had
the same number of twists—*one,* before cutting.
Let us make a Moebius strip that will lie flat, Fig.
29. There is one crossing of the edge with itself at
C, which cannot be removed by distortion, or even
the cutting and re-joining allowed by homeo-

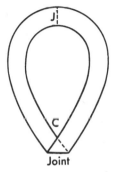

Made of 2 pieces
joined at J

Fig. 29

morphism. As was said earlier, a homeomorphic transformation allows us to make temporary cuts provided we reconnect what was connected, *point for point.* Thus a right-handed Moebius strip can be cut and re-joined into a left-handed one—even with more than 1 twist provided the number of twists is odd.

When Fig 29 is cut axially and opened slightly, Fig. 30, we see that the original twist shows in two places, *A* and *B,* imparting 2 twists to the new loop, but there is also a crossing of the loop with

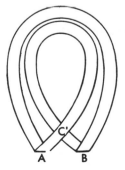

Fig. 30

95

itself at C', corresponding to the crossed edge C in Fig. 29. We now consider this new crossed part C', with the edges x and y emphasized, Figs. 31–33.

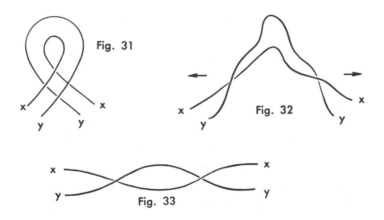

Fig. 31

Fig. 32

Fig. 33

We pull it out as per the arrows to find that we get 2 more twists, which added to the first two give a total of 4. It would seem that the original twist given to the Moebius strip was not alone: there was another concealed in the mere fact of the edges crossing one another—in making the new connectivity of the 2 edges of the untwisted and unjoined strip we used. The concealed twist seems to be apart from the physical act of twisting.

If we take paper strips and give them different numbers of twists, join them and cut along the axes, and count twists, we get the following table:

Before cutting	After
1	4
2	4
3	8
4	8
5	12
6	12
7	16
8	16
9	20
10	20 etc.

NOTE: When there is an odd number greater than one of twists before cutting, the resultant cut loop will have a knot in it, more complex as we increase the number. The significance of this will be explained later.

In each case, increasing an odd number of twists by one fails to increase the final number after cutting: this final number being the sum of the twists in the two loops that always result from cutting one that has an even number of twists. One would expect the number to be double that of the uncut loop, when the latter had an even number of twists, since cutting has merely given two new loops, both identical with the uncut version. Apparently adding 1 twist to an odd number is canceled by the fact of changing the connectivity of the edge back to what it was with the untwisted, original strip. So that at each increase from an odd to an even number we gain an addition of zero, but with an increase from an even to an odd number, we gain 4 new twists, as we did at the beginning in making our first twist, except that from one odd number of twists to the next odd number we have added 2 twists. This would seem to mean that in the odd numbers we do indeed progress according to the actual number of added twists—doubled by cutting—but with the persistent addition of the original extra two. Complicated, but logical. Now all we have to do is explain why the Gardner models keep giving 2 instead of 4 twists.

To return to the knots: the counting of twists after 3 and more odd-number twists is difficult until we recognize these facts. The mere twisting of one strip *around another* does not necessarily put a twist *individually* in either. Fig. 34 shows two

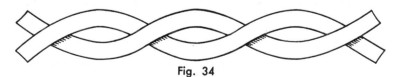

Fig. 34

paper strips cut with serpentine outlines: they are made to go over and under one another, *but neither is twisted in itself.* In Fig. 35 the same sort of thing has been done with a single strip, and to make this model—flat—we shall need two pieces of paper pasted together. It is obviously not twisted, to look at it, but looks are deceiving. It can be untwined from itself and will be as shown in Fig. 36, and we see at once that it is the same as

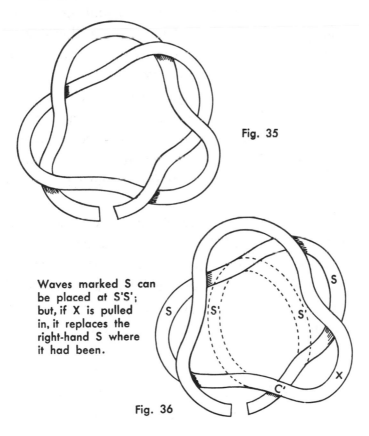

Fig. 35

Waves marked S can be placed at S'S'; but, if X is pulled in, it replaces the right-hand S where it had been.

Fig. 36

Figs. 31–33: disregarding the waves, it is a crossing of a loop with itself at *C*. When this is uncrossed as in Figs. 31–33, we find that it is in essence 2 twists.

All the knots generated by odd numbers of twists are of this form: no matter how many times the strip *twines* over and under itself, it makes two and only two circuits. Thus in counting we can hold one small section down with a thumbtack, trace the convolutions of the strip from end to end (from the thumbtack and back to it), keeping careful count of how many times it *turns over*—ignoring the number of times it goes over and under itself, and add two—that is, the number of twists it has—which in the case of Fig. 30 was two as it also made two circuits before being cut to open out.

Let us again examine the object we showed in Fig 13: it has 1 side, 1 edge, and no apparent twist. When cut axially it gives a loop with 2 twists— left-handed if opened to one side, right-handed if opened to the other. We can now see that it is truly a symmetrical Moebius strip, and the resultant 2 twists come from the mere fact of the connectivity of the edge: nothing more has been done which would add to that. We are finally getting somewhere. Fig. 37 shows again the cut version of the above, and it may not seem to have the two-circuit form mentioned earlier, but we turn one loop over into the position shown in Fig. 38, which now gives the double circuit. Since the intersection is imagined to be ideal, we can pull it out sideways according to the arrows, *or* in the opposite directions, getting right- or left-handed twists.

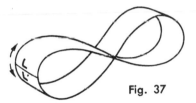

Fig. 37

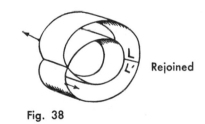

Rejoined

Fig. 38

If we take the "boned" model, Fig. 28, and cut it axially, after having joined only sections *A* and *A'*, we find that it is an orthodox Moebius strip cut axially: the result has 2 sides, 2 edges, and 4 twists. To count these it may be easier to cut off the unjoined arms. The twists will be right-handed, but if we had instead joined sections B and B', the twists would have been left-handed.

Let us re-examine the circular form of Gardner's model, Fig. 23, and do a little surgery: we find that it is very adaptable—even ambidextrous. By cutting on the dotted lines, Fig. 39, we get the homeomorph of the symmetrical Moebius strip

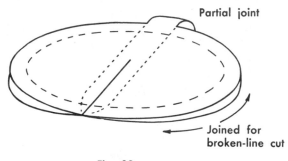

Partial joint

Joined for
broken-line cut

Fig. 39

of Fig. 13. Another version of the same thing is got by cutting on the broken lines, and both this and the previous one give 2-twist loops when cut axially. As we said of Fig. 23, cutting across just beyond the center C gives a disk (or cone) and a cross-cap: the word "beyond" is important.

The cross-cap obtained by cutting *through C,* the end of the intersection, looks like this, Fig. 40, and is the way it is shown in many books. As can be seen, it has 2 sides (shaded and unshaded, with

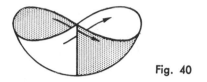

Fig. 40

arrows showing the connection), which is not correct. One should be able to cut a Moebius strip from it, but we cannot. If we extend the sides upward a little, the two surfaces are at once connected, Fig. 41, even when we choose the arbitrary shape shown here. The addition allows us to

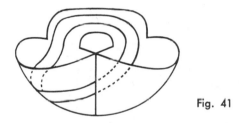

Fig. 41

trace out a left-handed Moebius strip, the inner line, and it is easy to see how a right-handed one could be got instead. This cross-cap has only 1 side, and fulfills the definition of being a projective plane with a point removed: in this case the point being enlarged to a disk. If we modify the shape of the part connecting the inside of the left with the outside of the right part, and the outside of the left with the inside of the right, we can cut a symmetrical Moebius strip from it, Fig. 42.

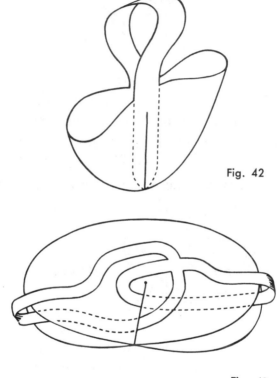

Fig. 42

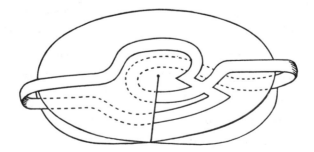

Fig. 43

The cruciform projective plane models—symmetrical, both arms left-, and both arms right-handed—can all be cut from the circular Gardner model. Fig. 43 shows the first and the last mentioned. We have only made the connections between the upper and under parts of the edges where necessary for the desired results, which shows that these can be got from a cross-cap also.

It cannot be too strongly emphasized that the Gardner model, like the ideal projective plane, is not symmetrical *because* of having "one pair of edges joined with a left-hand twist and the other pair with a right-hand twist"—when completed it has no twists, in spite of twists used in constructing the model. It is analogous to a circular track which can be said to "go around clockwise," if you traverse it clockwise, and with equal justice can be said to "go around counterclockwise" when so traversed.

On pages 101–102 the impression might be gained that when joining only sections B and B' (of Fig. 28) and cutting it axially we get the result of a left-hand twist *because* of a left-hand twist in the original "unboned" model. This is not so: we could have boned the same model differently, and the same two edges now have—or seem to have—a right-hand twist, and give right-hand twists when cut axially.

We show here two ways to bone the square and diagonal versions of the Gardner model, *all constructed the same way.*

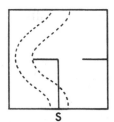

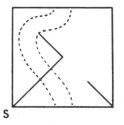

Giving left-hand-twisted Moebius strips

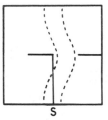

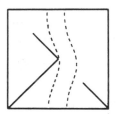

Giving right-hand-twisted Moebius strips

(Slits S must be rejoined.)

Fig. 44

This section has demonstrated how a small and unimportant-seeming anomaly in a lowly paper model—some people regard all models as lowly—

can lead to the understanding of a fundamental concept in topology: Symmetry. We can now understand how mathematicians think of a projective plane as being neither right- nor left-handed, and of a Moebius strip without twists. Perhaps the recently discovered left-handedness of certain subatomic particles will turn out, in the end, not to mean an asymmetrical universe, but merely that we happen to detect a left-handed section of something that is symmetrical in the Einsteinian space of more than three dimensions. Let's hope so.

7

Map Coloring

There are, in the early and more discernible reaches of topology, some theorems that have been propounded and proved. Some of them, like the Jordan curve theorem (pages 7–8), are famous possibly because they seem so obvious and are yet so difficult to prove.

Another of this kind is the tiling theorem. It states that on a (2-dimensional) surface it is impossible to make subdivisions—or cut it up into regions or faces—so that no more than two of them meet any point, *provided* none of these regions is larger in any direction than a chosen size. For example: if we divide the whole surface into checkerboard squares we can satisfy the last condition because if the chosen size is 1 mil, we make the square smaller than a mil (diagonally). But 4 squares come together at every corner, so the first condition is not met. (And 3 bricks meet at each corner on a brick wall.)

The first condition can be met by covering the whole surface with nontouching polka dots, but then the background—which is certainly one of the regions—would be bigger than the chosen size, if the chooser had his wits about him. Another way might seem to be making a series of concentric circles—they would never break the first rule by touching and therefore making points of meeting of more than two regions. Eventually, though, they would get too big, and even if they didn't, they would then leave too big a background region. You can't win. This would seem intuitively obvious but the proof is not. As a matter of fact the theorem goes on to generalize in an unlimited number of dimensions, but we shall not go into that. There is another theorem which is even more famous—or notorious—because after over a hundred years it has still not been proved.

This is known as the 4-color map problem—not officially a theorem until it is proved. It says that you need only 4 colors for a map—any map, provided it is drawn on a simply connected surface like the world, or this page.

As we all know, any number of regions—or countries—can meet at a point, but they are not considered as *all* touching so as to need different colors for every one. (A checkerboard can be, and is, colored with only 2 colors, although 4 squares meet at each corner of most of the squares.) The regions must have at least a finite amount of common boundary. Also we are not concerned with making the seas blue, or the British possessions pink; only with giving contiguous regions different colors. No one has ever made a map that needs more than 4 colors: no one has ever been able to prove that no such map can be made, and a great many learned brains have tried. It *has* been proved that 5 regions cannot be arranged so that each one touches every other one—the proof is trickier than it seems, if it is made absolutely rigorous—but that is not the same thing.

We can start drawing a map, and color it as we go, and eventually get ourselves stuck and have to backtrack, and rearrange previous colors. It has always been possible to do this successfully, but so far no one has proved that this must be possible.

The most irritating part is that a much harder-seeming theorem has been proved: that on a torus, or any doubly connected surface, 7 colors may be needed, and 7 colors are always sufficient. In case the reader likes to puzzle people, Fig. 1 can be drawn on a paper torus (doughnuts are no good

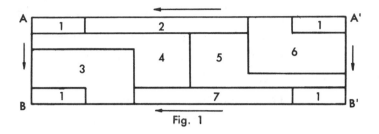

Fig. 1

for cartography), and offered for coloring after appropriate palaver. As can be seen, there are 7 regions, and each touches every other one (remembering the joints as shown.)

It should be explained that the regions on opposite sides which meet along a *fold* must have different colors. Even the Moebius surface has been proved to need no more than 6 colors, and in fact sometimes needs that many. If a strip is subdivided as in Fig 2, then twisted and joined, we see that there are 6 regions, and each touches every other. The strip having one side, we regard it as transparent: the colors show from both *directions* (see *orientability,* pages 25–28).

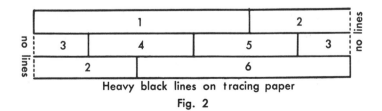

Heavy black lines on tracing paper

Fig. 2

The 4-color problem has been attacked from various angles, the most promising-seeming being Euler's formula for polyhedra, since a map can be topologically changed to one, and the formula applies anyway, as we have seen (page 17), to any figure that consists of faces (regions), edges (boundaries), and vertices (meeting points of boundaries). In spite of exhaustive analysis the basic problem has not been solved, although some interesting theorems have appeared as by-products. In a sense we could call it the 3-color problem, because if we can draw a map so that the outside regions need more than *3* colors, we can then surround it with a single region which will need a fifth color.

This does not mean that this map uses only 3 colors, except for the surrounding region, to come out right, but that in all cases we must be able to *rearrange* the colors so that only 3 are needed for the exterior regions. In this map, Fig. 3, we start coloring the inner regions: 1, *2*, and 3. Then the surrounding ones: 1, *2*, and 3 as shown, but *x* will need a fourth color, and *y* will need a fifth. To avoid this we must "retire" the fourth color at *x* to one of the inner regions, which lets us use 3 for *x*. If we can find an infallible method of retiring the exterior fourth colors at *each succeeding stage*

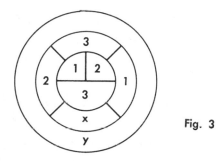

Fig. 3

as we move outward, we shall have solved this phase of the problem.

Any map can be made more uniform by changing it into what is called a regular map—one in which no more than 3 regions meet at any point. This will not affect the coloring, because all we do is to say that some nontouching regions do touch —and if we color them differently no harm is done. The usual method is to put a new area, *A* (Fig. 4), in place of the point of meeting of more than 3 areas, *p*. We now have the four points *a*, *b*, *c*, and *d*, each being the meeting of only 3 areas (regions). If we color this second map correctly, and then remove *A*, the result will still be cor-

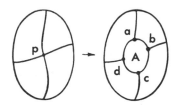

Fig. 4

rectly colored, albeit perhaps with 3 colors when 2 would have done as well. We have sacrificed simplicity for uniformity—sometimes useful in mathematics.

We can get a still greater uniformity if we first convert the map to its dual. This is a connectivity network, or grid, in which the areas are represented as points, and the connections—the touching of areas—as lines joining the points. The map shown in Fig. 3 is given in Fig. 5 with a point in each area, and a line connecting them in every case of touching. Then the map is removed, leaving its dual (Fig. 6). Since it also happens that the map was regular, all the polygons of the grid are

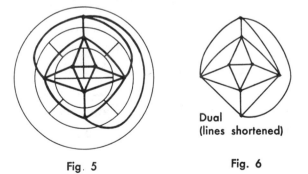

Dual
(lines shortened)

Fig. 5   Fig. 6

triangles. Had it not been regular we could have made it so without the addition of new areas, as in Fig. 4, but by triangulating. The map in Fig. 7

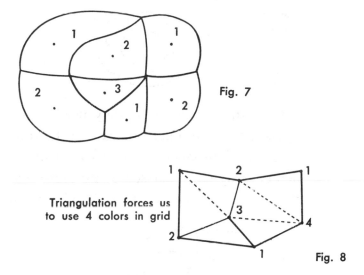

Fig. 7

Triangulation forces us
to use 4 colors in grid

Fig. 8

gives the grid in Fig. 8 and the square and pentagon are made into triangles by adding the dotted
lines, which will not hurt the solution for the reasons given above, page 113.

Coloring means, here, labeling the points so that
no two that are directly connected have the same
number-label. It will be found that we are no
nearer the solution, but at least we know we are
dealing exclusively with triangles. We can now
simplify further. It is obvious that in a map any
islands, or enclaves, can be ignored since they
only touch the area around them, and can be colored—labeled—differently from it. Also, an area
surrounded by 2 or 3 mutually touching areas

115

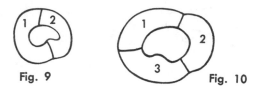

Fig. 9   Fig. 10

(Figs. 9–10) can be ignored as they need 2 or 3 colors anyway, and the inner one can be the third or fourth color. This also applies to a *group* of areas when entirely encompassed by no more than 3 mutually touching areas (Fig. 11). The group

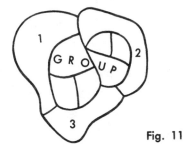

Fig. 11

constitutes a *separate* map, in that its personal problems cannot affect what lies outside the encompassing areas: the latter need 3 different colors no matter what, and if we prove the theorem for any map, it will also apply to the isolated group. This means that no grid will ever have a triangle containing any other point or line, except the *exterior triangle* if there is one. Also, we shall never have less than 4 lines meeting at any point (Figs.

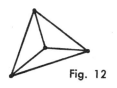

Fig. 12

Fig. 13

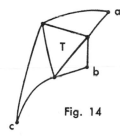

Fig. 14

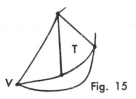

Fig. 15

12–13), as it could only occur in a single (possibly distorted) triangle—forbidden in Fig 10. Every triangle T would have three other triangles on its sides, with *independent vertices a, b,* and *c* (Fig. 14), since if two of them have a common vertex V, the result would be Fig. 15 (the grid of Fig. 10). Lastly, all redundant connections are deleted, as we would color the connected points differently anyway.

If the map is on a sphere we open it out flat, as we did with the polyhedra on pages 11 and 12, making the area on the back into the exterior area of our new map. This, in the grid—or dual—becomes a single exterior point p, which is connected by a line to each of the previous exterior points, and

117

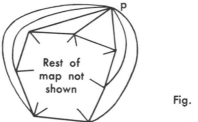

Fig. 16

will always give a triangle for the exterior polygon (Fig. 16).

The reader is now on his own: since the ground has been so thoroughly gone over before, no inducement is offered beyond entertainment. But it may be that a fresh, more intuitive eye might see what the experts have missed.

As a means of getting the feel of the difficulties you run into when coloring maps, this 2-handed game helps and may even amuse: Player A draws a region. Player B colors (labels) it, and draws a new one. Player A colors it and adds a third. This goes on until somebody gets stuck and has to use a fifth color. Traps can be set for the unwary—and sometimes foreseen and avoided.

PUZZLE: You are required to color the map (Fig. 17). Each region has an area of 8 square feet except the top one, which has 16. You have the following colors: exactly enough RED for

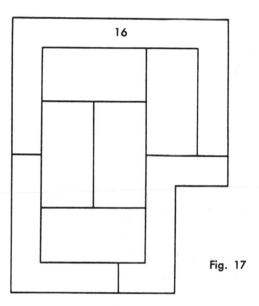

16

Fig. 17

24 square feet, enough YELLOW for 24 square feet, enough GREEN for 16 square feet, and enough BLUE for 8 square feet. The rules are as usual: no touching areas may be colored the same. Watch out for unicorns. (Answer in Appendix.)

8

Networks

The Koenigsberg Bridges

In the last chapter we changed a map into a network to be able to see what was what. As it happens, this was one of the beginnings of topology. At Koenigsberg on the river Pregel—then in East Prussia, now in the U.S.S.R.—there used to be seven bridges connecting two islands with one another and the banks, as in Fig. 1. In the early eighteenth-century they had a puzzle which was to see if one could walk a complete circuit of all the bridges—once and only once each.

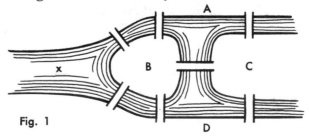

Fig. 1

It turned out to be like the 4-color problem, because no one could do it, though no one could prove that it was impossible. The reader can try on the diagram, and see. In 1736 Euler (page 10) proved that it was impossible—by reducing the map to a network (Fig. 2) in which the land areas are rep-

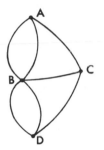

Fig. 2

resented by points and the bridges by lines. He worked out the general law for all such networks, as part of his investigation of polyhedra. It can be seen that unlike our networks for maps it has redundant connections at *A–B* and *B–D,* but that is where the bridges happened to be.

The law is as follows: at each vertex a certain number of lines (bridges in this case) meet. If an odd number, it is called an odd vertex; if an even number, an even vertex. Now, it can be proved that there can only be an *even number of odd vertices,* or *none.* (We suggest that the reader try proving this, using the proof of the other Euler law on pages 15–17 as a guide.) A circuit where each line is traversed only once can only be made when there are either *no* odd vertices, or 2. Koenigsberg consisted of 4 odd vertices, and the circuit is impossible. Do not go there to try this—they've built a new bridge at *X,* Fig. 1, which changes the north and south banks into even vertices. It will be found that to make the circuit we have to start at an odd vertex.

A completely unfair bet can be made on the basis of this law: ask someone to draw any network and then bet him whether it can be traversed or not. First, of course, you underhandedly count the number of odd vertices, and bet accordingly. Most of us knew a simple version of this when we were kids: Fig. 3 cannot be traced in one line, but Fig. 4 can, *provided* you start at *A* or *B.*

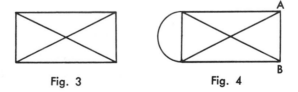

Fig. 3   Fig. 4

The general law says that the number of disconnected trips needed to make a complete traverse is half the number of odd vertices (always an even number).

Betti Numbers

We can construct a network that has no enclosures or loops: a network that is all in one piece, in which every line terminates in an otherwise free vertex. This is called a *tree,* and it is fairly easy to see that there will always be one more vertex than there are lines, or edges, as we called them in Chapter 1. For one thing, if we apply the Euler formula for *any* figure, since a tree has only one face—the background—and the formula is $F-E+V=2$, we get $1-E+V=2$, or $V=E+1$.

We can make any network into a tree—i.e., leaving one connected figure—by removing some of the lines, or edges (Fig. 5). Say we have to remove B edges (in this case $B=2$), to get rid of all enclosures: we had E edges and V vertices to start with, and we have just seen that in a tree $V=E+1$, so we now have $V=1+E-B$, or $B=1+E-V$—in this case $1+8-7$, or 2. B

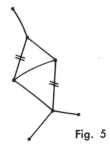

Fig. 5

is called the Betti number of a network, and it is always equal to the number of faces minus 1. (It is named for Enrico Betti, nineteenth-century Italian mathematical physicist.)

The same will obviously be true for polyhedra, as in the tetrahedron of Fig. 6: 6 edges, 4 vertices, and 4 faces. We can remove 3 edges without cast-

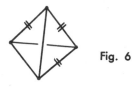

Fig. 6

ing any vertex adrift: so $B=3$, or one less than $4F$. A number that is merely $F-1$ may seem a bit trivial, but it is basic to the idea of connectivity, and *it applies also to surfaces,* though in a different form. A disk—or anything topologically a rectangle without a hole—cannot be given any cross-cut without dividing it into two pieces. (A cross-cut means a cut that starts at an edge and ends at an edge.) This means that a (topological) disk has a Betti number of zero. On the other hand an annulus, or a Moebius strip, has a Betti number of 1: we can make the dotted-line cuts in each of these (Figs. 7–8) without dismembering them. A disk with 2 holes would have a Betti number of 2. How about a sphere? Since it has no edge we can-

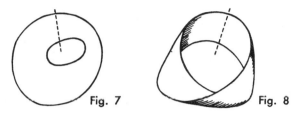

Fig. 7   Fig. 8

not make a cross-cut, and if we make a hole to get an edge, we have a topological disk—stretch the hole and flatten it all out.

But a torus, or a Klein bottle, is something else again: a hole does not make a torus into a surface we can deform into a disk or even a cylinder, and the same is true of a Klein bottle. Of course by a hole we mean a mere puncture—not punching a small disk out like a train ticket. (The deformations of a punctured torus will be examined in the next chapter.)

It is fairly obvious that there are two kinds of cross-cut we can make in a punctured torus that will not divide it: one going around the tube part —cylindrical—*x*, and the other around the center —annular—*y* (Fig. 9). When *x* and *y* are cut, the

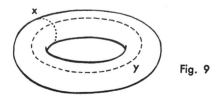

Fig. 9

torus remains in one piece, and it can be seen on pages 71 ff. that the same holds true for a Klein bottle. Both these surfaces have a Betti number of 2.

To find the Betti number without making a hole, we make a loop-cut, which is just what it sounds like: it starts at any point on the surface, and returns to it without crossing itself—a Jordan curve. Here the formula changes slightly: a loop-cut in a disk divides it in two, but so it does with an annulus or cylinder. So we count the number of edges and say that B equals the number of loop-cuts we can make in a surface *without dividing it into more pieces than there were edges*. Thus a disk has 1 edge, and no cuts can be made without getting more than one piece, so $B=0$. An annulus has 2 edges, and one loop-cut *can* be made without getting more than *two* pieces: $B=1$.

But a Moebius strip has only one edge, and one cut can be made—along its length—without dividing it, so it has $B=1$ also. With a nonpunctured torus we can make two cuts—parallel to the ones we made before—so $B=2$. Same for the Klein bottle but not, as it happens, for the projective plane. The latter is very hard to visualize, but it can be shown if a version of Gardner's model, Fig. 1 (page 79), is cut along both axes: it is divided in two—a cone and a cross-cap. It is much easier to use the admittedly incomplete model of Fig. 3 (page 51), but start with the one of the Klein bottle (Fig. 2) first—it is most astonishing when cut along the dotted lines (both shown in Fig. 10): it becomes a flat hollow square. The projective plane, on the other hand, comes apart. It is not the resultant shapes that matter, but their number: 1 for the Klein bottle, and 2 for the projective plane, so for the latter the Betti number is 1, like the Moebius strip. It seems rather unfair.

All this does not mean that *any* loop-cut can be

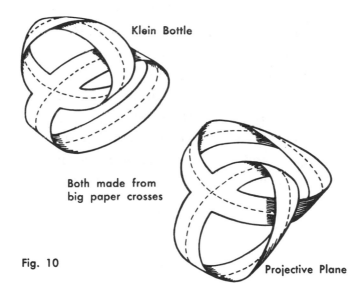

Klein Bottle

Both made from
big paper crosses

Fig. 10

Projective Plane

made in a *B2* surface with impunity: it could be a small one in the side that removed a round piece. Also some cross-cuts will divide a Moebius strip: C-shaped, cutting the outer edge twice; but the point is that there are cuts that will not divide it. Another fact is that if we mark the permissible—nondividing—loop-cuts and cross-cuts, puncturing the surface for the former, we find that each loop intersects one cross-cut mark. This "duality," as it is called, has been proven to persist in any number of dimensions, by S. Lefschetz of Princeton University (1927), and is a strong kind of invariance. For now we can say that (1) the number of edges, (2) the number of sides, and (3) the Betti number are invariants for 2-dimensional surfaces. The Betti number for a cat is 9.

Later on we shall be discussing series, and while we are on the subject of cuts, this seems a good moment to pose the following problem. If we cut a piece of paper in two, put the pieces together, cut again, and so on, it is obvious we first get 2, then 4, then 8 . . . doubling every time. Parenthetically, and not topologically, if this were done with a playing card a mere 52 times the resultant pile would reach far beyond the sun. The numbers 2, 4, 8, 16 . . . etc., in this case form a geometrical series.

So far so good, but what would happen if we folded, and then after, say, 6 folds we cut? The folding can be done two ways: each time at right angles to the previous fold, or all parallel. Also the cut can be at right angles to the last fold, or parallel. With parallel folds and right-angle cut the answer is jejune: always 2 pieces; but with a parallel cut things get complicated. At the first fold we get Fig. 11, shown in diagrammatic cross sec-

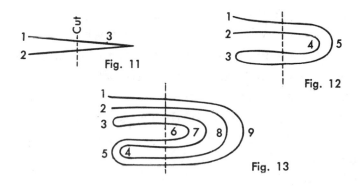

Fig. 11

Fig. 12

Fig. 13

tion. Then Fig. 12, the second fold, and Fig. 13, the third, and so on. If we count the pieces each time we see the beginning of a series: 1—with no cuts (not shown), 3, 5, 9. . . . The suspicious minded may notice an odd similarity to the series got by cuts alone, except that here 1 has been added to each term: $0+1, 2+1, 4+1, 8+1$. . . .

Experiments in Topology

At this point paper and scissors seem simpler than diagrams. Sure enough, the next number is 17. If Figs. 11–13 are examined we can see that what happens is the same as doubling each time, except for the 2 loose, or unattached, edges: if these were joined we would get at that place 1 instead of 2 pieces. This is what adds 1 to every term. With right-angle folds and either kind of final cut, the series progresses at half speed: 2, 3, 3, 5, 5, 9, 9 . . . in the case of the right-angle cut, and starting at once with 3, 3, with the parallel cut. Experiment with paper will show why, but not for long: it gets too thick to cut, rather quickly. Again parenthetically, it is a safe bet to challenge anyone just to fold a piece of paper—any piece —ten times: first in half, then crosswise and so on. It is totally impossible . . . they don't sell paper big and/or thin enough. We now come to the crux of the matter:

We fold and cut every time, *but not all the way through* until the final cut. How many pieces do we get for each number of partial cuts? What is the series? Because of increasing thickness we shan't be able to carry out experiments with actual paper beyond a mere 6 or 7 folds, but attention to what is happening will suggest Fig. 14: x equals final cut, the others are numbered: all double alter-

nately. We shall confine the way of folding to right
angles every time as otherwise things get ridicu-
lous—it's bad enough the right-angle way. There

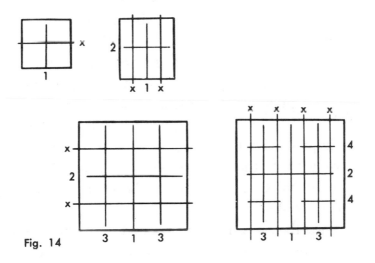

Fig. 14

are 4 directions to cut in: north, south, east, and
west—east and west being the same thing really.
It will be found that the direction makes little difference. (When making a cut don't go too near
the edge.)

The folds, of course, involve only the small uncut part. A hint without which the problem is unfair: sometimes one finds there are two or more
series superimposed, or alternating, where one applies to the odd and the other to the even numbers
of operations. Answer in the Appendix.

131

Experiments in Topology

Knots

The subject of knots has been taken up by topologists, but not very much has been proved beyond saying that a knot cannot exist in more than 3 dimensions. This is reminiscent of the fact that a Jordan curve only divides a surface—not 3-dimensional space. *O* divides this page into two parts: inside and outside it; but a loop of string hanging in the middle of the room doesn't divide anything. Perhaps it would be more apposite to say in discussing knots that if two loops were linked in 3-dimensional space, they would be separable in 4-dimensional space.

Nevertheless, a repetitive sound in mathematics is usually worthy of investigation: there are no coincidences in mathematics—only some unfortunate approximations, like π and 3. (This proximity has misled many people: the Bible flatly says that π *is* 3, and there are still some cranks who believe it. A crank is an unsuccessful eccentric.) The difficulties in classifying knots have not been overcome. For one thing it is not enough to say that two loops are linked merely on the grounds that they cannot be separated. The two annuli of Fig. 15 are linked and cannot be separated, but how about the next three (Fig. 16)? None of them

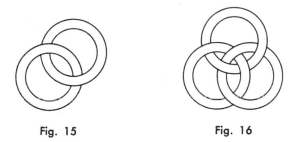

Fig. 15   Fig. 16

is linked with another, and yet they are unseparable.

Some topologists aver that all knots are intrinsically the same—just loops or circles—but it would be hard to convince a sailor or a boy scout that the mere joining of the free ends of a knotted rope affected the knot, or that a granny was intrinsically a bowline. The most effective thing about a knot is its holding: but that brings in friction, which is outside our subject. Knots occur as a sort of by-product of surfaces: the edge of a Moebius strip is a twisted loop with regard to 3-dimensional space, and can be topologically deformed into a circle, as on pages 80–81. But if the strip has been given not 1 but 3 half-twists before joining, it would be in the form of a trefoil knot (Fig. 17). When cut, and the ends separated, it turns out to be the simplest of all knots. An old conjuring trick was based on this: the performer prepares 4 large paper bands: one untwisted, one

133

Fig. 17

with 1 half-twist, one with 2, and one with 3. When these are cut down the center, the first gives 2 unlinked bands; the second, 1 long band; the third, 2 linked bands; and the last, 1 band with a knot in it, the latter being the trefoil. Trefoils can be right-handed or left-handed, but, though the difference is obvious, it is impossible to define which is which without reference to some existing standard.*

It is instructive to see how dictionaries define "left" and "right." The *American College Dictionary* says that left is toward the west when facing north, and right is toward the east, etc. *Funk & Wagnall's* is more poetic: for left, "toward the north when facing the sunrise." One might hope that for right they would change it to sunset, but alas, no: south for north. *Webster's New Collegiate* is anthropocentric: the left side in man is muscularly the weaker, while the right is more

---

* It is conceivable that such a standard may exist universally—in the atom—but it would not be topologically definable.

skillful. The *Encyclopaedia Britannica* is cagey: they stick to political differences. It is like trying to decide which is width and which is thickness— it depends how you hold your head. It is a relative matter. A long, useless, and entertaining discussion can be promoted by asking someone—preferably a dogmatic type—why it is that mirrors reverse you left-to-right, but not top-to-bottom. If he is up on this one, he may say that it does neither —it reverses you back-to-front (true). Then ask him why this makes a man-holding-a-pen-in-his-right-hand into a ditto-*left*-hand, but *still upright*. He will probably get cross.

9

The Trial of the Punctured Torus

REFEREE: This hearing will come to order. Please state your case, Mr. Jones. There is a blackboard if you want to make any drawings.

JONES: My worthy opponent, Dr. Situs, claims he has turned an inner tube inside-out—

SITUS (*interrupting*): After cutting a small hole in the side.

JONES: Right. He has boasted of this as something surprising, and I would be surprised, too, if the claim were true. My contention is that it is only half inside-out (Fig. 1); just look at it.

Exhibit A

Fig. 1

The Trial of the Punctured Torus

REFEREE: Dr. Situs, will you state your case?

SITUS: My opponent is evidently misled by the irrelevant fact of its present outline, like that of a hollow handle removed from a teacup. He evidently does not recognize a torus when he sees one. This one is distorted and bent, but topologically it is a torus, though punctured, as it was before inverting it.

REFEREE: Inverting?

SITUS: Turning it inside-out—but the other sounds more mathematical. My point is if the exhibit is examined we see that it is merely a torus with a long, narrow, and bent hole through the center: what would be called the hole in the doughnut. It has a flattened and elongated inner space where the compressed air would normally be. I can show on the blackboard how its present shape is deformed into a more recognizable torus without tearing or joining. (*Goes to the blackboard and draws Fig. 2.*) As you see, all we have done is to shorten it a little.

Fig. 2

JONES: And very conveniently reduce the inner, unturned part to a thin ring!

SITUS (*brushes this aside*): I should also like to point out that the inner tube I started with was gray: it is now black! This is because what was the inner surface was black, and now *all the exterior is black.*

REFEREE: May we have a demonstration of this?

SITUS (*unhappily*): Oh . . . well, if you insist. (*Shrugs: he struggles with Fig. 1 for a while; there is a resemblance to the Laocoön statute and finally the tire is back in its original form [Fig. 3] —gray all over.*) There! I shall now explain this by a series of diagrams. (*Goes to the board.*) In Figs. 4 and 5 we begin to stretch the hole until it

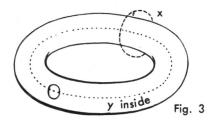

Fig. 3

Fig. 4

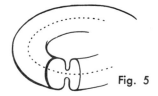

Fig. 5

(Ignore the dotted lines — they will be referred to later)

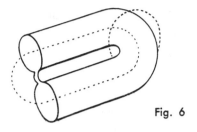

Fig. 6

leaves only a narrow connection, Fig. 6. Then, Fig. 7, we begin to roll part of the torus back on itself like the top of a stocking, labeling the newly appeared part of the inner surface *P*, and the outer surface *Q*. Then we continue as in Figs. 8–9 until *P* is back next to *Q*, reducing the hole to its orig-

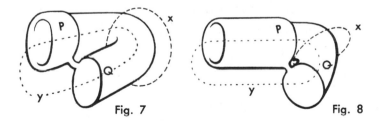

Fig. 7

Fig. 8

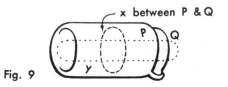

Fig. 9

inal width. Fig. 10 shows the end view, and we see that the hole is now in the form of a doubly bent Jordan curve, which we shrink again, Figs. 11–12.

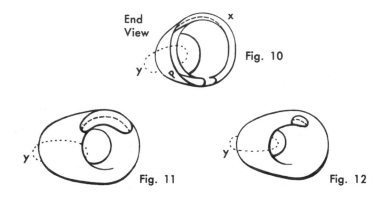

End View

x

y

P

Fig. 10

y

Fig. 11

y

Fig. 12

In Fig. 13 we shorten the whole torus back to its doughnut shape. It will be noticed that *P*, originally on the inner surface, is now on the outer, while the reverse has happened with *Q*. The torus is inside-out.

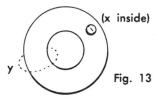

(x inside)

y

Fig. 13

JONES: I admit the line of reasoning up to Fig. 9, but you should have stopped there because that's as far as you went with the real inner tube. Let me redraw Fig. 9 to conform to reality—or your idea of it. It looks like this (Fig. 14), and the

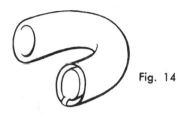

Fig. 14

inner half is exactly as it was at the beginning: you merely *call* it inverted. All the points on it that were originally facing toward one another are still facing toward one another, while those that faced away—

SITUS (*interrupting*): Yes, but the first ones you mentioned faced toward the *inner* space: they now face the *outer*.

JONES (*goes to board*): I think I can clear this up. Here is a glove (Fig. 15); let us say black inside and gray on the outside. We now proceed to turn it inside-out—first by turning back the wrist as far as we can (Fig. 16). We notice that the

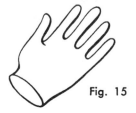

Fig. 15

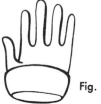

Fig. 16

141

thumb and fingers are still as they were (Fig. 17). Now we push them out, Fig. 18, all except the *middle finger*. Note, gentlemen, that if we seal off

Fig. 17

Fig. 18

the wrist hole, all of what the learned doctor calls the new "exterior" is now *black:* but what about the middle finger? Is it inside-out, or are we merely playing with words?

SITUS: But that isn't a torus!

JONES (*craftily*): My worthy opponent is fond of showing topological equivalences: may I suggest that a glove—when we seal the wrist hole—is a topological sphere? Suppose we start with a one-fingered glove (Fig. 19), and we sew the tip to the opening: Now we begin to invert, as the Doctor would say. We continue, but lo and behold we can never turn more than half of it inside-out, even if

Fig. 19

we move the tip back into the inside. Were there a small hole in the tip we would have an exact homeomorph of a punctured inner tube.

SITUS: And it would be inverted!

JONES: May I ask you to specify the difference between the condition of the middle finger, Fig. 18, and that of the others?

SITUS: It's, er . . . This is absurd! Ah! I have it! (*Goes to board.*) Here (Fig. 20) is a hollow punctured sphere: we turn it inside-out through the hole—surely you will allow this possibility? I

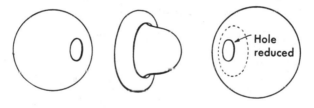

Fig. 20

now reseal the hole and push in the side (Fig. 21) —does this mysteriously, not to say semantically, cancel half of the inversion?

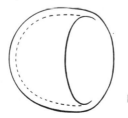

Fig. 21

JONES: In this case, no. But we were discussing tori.

SITUS: May I ask how a torus can be any more inside-out than mine? It has all of its once-inner surface facing the exterior.

JONES: Allow me. (*Takes inner tube, and produces scissors and quick-drying cement.*) I cut through, Fig. 22, I invert the resultant cylinder, Figs. 23–25, *all* of it, and rejoin the cut ends, Fig. 26. I think we can all agree that this is what would properly be called an inverted torus. It looks like the original one; not like a hollow teacup handle.

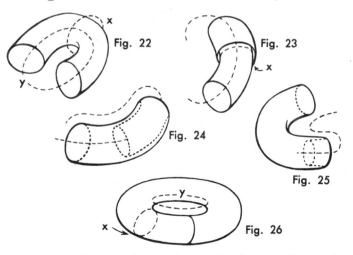

Fig. 22

Fig. 23

Fig. 24

Fig. 25

Fig. 26

SITUS: You can't go by looks in topology: by what standard is your torus more truly inverted than mine?

The Trial of the Punctured Torus

JONES: Let us return to your own diagrams. I should like to refer you to a topological invariant: linkedness—

SITUS: Hey! That's not—

REFEREE: Later, Doctor, later. Proceed, Mr. Jones.

JONES: Two closed curves that are linked cannot be unlinked by any topological deformation. I now add two such curves to your diagram (pages 138–140) : one, x, running around the tubular part and one, y, running through it. Observe they are linked. I draw them in each successive stage of your so-called inversion, and at the end they are still linked!

SITUS: What difference does that make?

JONES: In *my* one they get unlinked. (*Demonstrates,* Figs. 22–26, the dotted lines.)

SITUS: That's not fair! You cut one of the loops . . . (*He abandons this tack as he realizes that even if the loops had been cut in his inversion, the linkedness would not be altered.*) The point is trivial: linkedness is not an invariant through n dimensions. Everyone knows that a knot cannot exist in more than 3 dimensions for this very reason!

JONES: May we return to my opponent's main claim: that all the inner surface was made to face the outer space? This is the same as saying that if we imagine a series of lines that project from the surface in question, as in this cross section of the inner tube (Fig. 27), according to the Doctor they reach into—have their neighborhood in—the inner space. He performs his deformation and they now reach into the outer space (Fig. 28). But I hope he

Fig. 27   Fig. 28

will recall that Fig. 29 is topologically identical with Fig. 30, and the addition of another dimension permits this deformation.

Fig. 29   Fig. 30

SITUS: But I did not make use of a fourth dimension!

JONES: Neither did I, to unlink the loops.

SITUS: Maybe not, but you implied the existence of those little lines drawn radially from the surface earlier in your own argument. When you spoke of half my torus being "exactly as it was in the be-

146

ginning," you were really saying that the lines that pointed towards one another before inversion did so afterward. By your own admission, that is topologically trivial. Furthermore, if the torus has a grain running in the annular direction like this (Fig. 31), we find that after *my* inversion it runs the other way (Fig. 32), and it is of course on both surfaces. That is a real change!

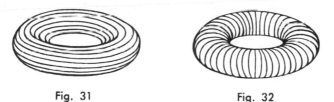

Fig. 31   Fig. 32

JONES: I fail to see what it has to do with the price of eggs. In what way is it a property of, or an indication of, inside-outness? I should say it shows that your distortion was of another kind altogether! (*They stare at each other, breathing hard.*)

REFEREE: I shall now sum up. On Dr. Situs's side: (1) He brought the inner surface to face the outer space, and I'm inclined to say that that is the usual meaning of "inside-out"; (2) he began and ended with a torus, topologically speaking; and (3), for whatever it is worth, he changed the "grain" from cylindrical to annular. On the other hand, Mr. Jones has (1) also brought the inner surface to the outside, and (2) he also began and ended with a torus, and (3) while he did not alter the grain, he unlinked the linked curves left unaffected by Dr. Situs. The jury will please consider their verdict. (*They do so, and perhaps the reader will join them and help decide what the verdict should be. The writer has found that discussion of this question can fill an evening—it is in one sense a matter of definition, but that does not make it trivial, as the good Doctor might say.*)

10

Continuity and Discreteness

The "Next Number"

Hairsplitting, Webster notwithstanding, is not the same as quibbling. Topology prefers to be noncommittal as to exact size and shape, but is most careful about exact meaning. We feel we know what we mean by a continuous line: that it has no gaps—but the definition needs improving. A string with a gap would be in two pieces, but how about a net? Then for example we might think of the list of whole numbers between 1 and 20, and in a sense if 13 were missing we might say there was a gap, but if not, not. But if the numbers corresponded to the marks on a ruler—admittedly incomplete as they rarely are finer than $\frac{1}{32}$ inch—how do we make the series of numbers continuous? What would it mean? Can we keep filling the smaller fractions in between, and then between them, forever without end? If there were no end to this process—truly no point at which we could say the numbers were all there—we could never do it. Yet if we say that there is such an end, how can we say that we cannot put yet more, still smaller fractions between the last ones?

Perhaps instead we could start at one end with a point, and put one next to it, and so on forever. But if a point, as the geometry books used to say, has position but no magnitude—no size in any direction—to say that another point, also having no size, was *next it* would imply that there was *no distance between them.* This would mean that their positions were the same, and since the only thing that distinguishes one from the other is its position, it would mean that they were the same point: we have consequently failed to put down a *next point.* This is hairsplitting with a vengeance, and important in topology.

A similar state of affairs is found with maxima and minima. If we define a set of all the fractional numbers from 1 to 2 *inclusive,* then 1 is the minimum and 2 is the maximum. But if we had said "greater than 1 but less than 2," things change. The minimum would be the next fraction above 1, and the maximum would be the next fraction below 2—and neither can be named. (The first example is what is called a closed set of numbers, and the second an open set. Unimportant as the distinction may seem, it will be seen later that it is useful in reasoning about sets of things—especially points.) But what has a discussion about numbers to do with topology?

To explain this we shall have to examine what the idea of a *set* of anything can mean. We shall see that certain generalizations and definitions can be made about sets that cannot meaningfully be made about their members individually. This is reminiscent of the way topology ignores individual differences among, say, figures bounded by closed curves, and treats them as a group that have certain invariants in common. It is no coincidence.

Continuity

When something is not continuous we call it discrete. The list of all the integers is discrete—though infinite; the sands of the beach are discrete, and so is water if we consider the molecules. It is not enough to say a line is continuous *because* it has an infinite number of points: there are an infinite number of rational fractions in it, but what about the infinite number of gaps always left to put more in? "Rational," of course, means any fractional number made by dividing one whole number by another: n/m. Any conceivable rational number (8 is $\frac{8}{1}$) can be so written, but the irrationals, like π or $\sqrt{2}$, cannot—they can only be approximated.

As a rule topologists confine the use of the word "continuous" to processes, rather than spaces—a line is a 1-dimensional space—but if we must use it for a line, then *continuity relates the set of all the points on the line to the set of all the real numbers*. Real numbers means the rational *and* irrational ones. It has been proved that there are at least as many (infinite) of these irrational as rational numbers. The main difference here is that the rational numbers are denumerable, while the irrational are not. It means that they can be counted off: 1, 2, 3, etc. to ∞; can be, in the sense that we know how to do it without missing any. In the case of all the rational fractions—n/m—though we cannot count them in their natural sequence (since we can never say what the next one is), there is an ingenious dodge to get around this.

All whole numbers are written as fractions: $1 = \frac{1}{1}$, $2 = \frac{2}{1}$, etc., and the rest are reduced to their lowest terms, as otherwise we would keep redundantly listing the same one: $\frac{12}{10}$ for $\frac{6}{5}$, etc.

We now start at $1/1$, then $1/2$, then $2/1$, $1/3$, $3/1$, $1/4$, $2/3$, $3/2$, $4/1$, $1/5$ . . . , this method being to increase not the value of the fraction as we go along, but *the sum of the integers used in the fraction.* Thus $1/1$ obviously has the smallest sum; $1/2$ and $2/1$ have the next—we put them down as we come to them in their actual order of value—then $1/3$ and $3/1$ (adding up to 4 each); then $1/4$, $2/3$, and $3/2$ (adding up to 5), and so on. As we go on, the missing ones, like the fractions between $1/4$ and $1/3$, get more and more filled in. We never get to the end any more than we would counting off 1, 2, 3 . . . etc., but we know that we have the fractions arranged in a new but logical way in which all the rational numbers will be turning up *only once each.* This means that we can label each uniquely as the 1st, 2nd, 3rd, and so forth.

This list shows the rational numbers and under them the sums of their numerators and denominators, and under these their rank:

$1/1$	$1/2$	$2/1$	$1/3$	$3/1$	$1/4$	$2/3$	$3/2$	$4/1$	$1/5$	. . .
2	3	3	4	4	5	5	5	5	6	. . .
1	2	3	4	5	6	7	8	9	10	. . .

Any set of numbers—or points, or in fact anything at all—which can be counted off (denumerable) is discrete. The favored word to describe a nondenumerable set—like all the real numbers, or all the points in a line—is a *continuum*—rather than *continuous,* as that usually refers to a process. A variable number passes through all the points in a line: the passage is continuous.

So we see that there are two ways of pointing to infinity: 1, 2, 3 . . . to infinity; or, all the points in a segment of line. More important is that in each of these there are two kinds of infinity: denumerable, e.g., rational numbers, and nondenumerable, e.g., all the points on a line.

Neighborhoods

The reader may remember that on page 146, Mr. Jones spoke of some lines having their "neighborhood in the inner space." We saw what he meant —just that that's where they were—in a sense rather vague. Yet with all this hairsplitting precision about points and infinity and so on, it is also useful to have a sort of controlled imprecision about certain things. For example, on page 108 we mentioned the tiling theorem, and referred to regions being no larger "than a chosen size." This was pretty vague, but its implication was that the tiling theorem applied, no matter how small the chosen size was. This is a way of getting around saying that "it has *some* size" and at the same time "it is infinitesimal—or of no size"—a falsehood, and imprecise as the dickens. The other is at least not a contradiction.

Similarly it is useful to have a way of talking about one point, in a given space, being "sufficiently near" another point. This is meant in the same spirit as "chosen" in the tiling theorem: it means "as close as you please." To refer to the *neighborhood* of a point does not say how big the neighborhood is, but only that it contains the point, and contains it in such a way that within the neighborhood we may put another point as close to it as we please. This is quite imprecise as to how far away another point can be: though in a measurable—metric—space we may use the symbol ε (epsilon) to label the diameter of the neighborhood, and yet not commit ourselves as to the size of ε. This ε-neighborhood of point P means that everything in it is less than ε away from P. Note that we do not say "ε or less" but only "less," which means that there is no maximum (see the discussion of maxima and minima on page 150).

Imprecise as all this is, it allows some fine hair-splitting. Suppose in a line we choose a segment S

(Fig. 1). There is a point P in S, and it has a neighborhood N. One thing is clear: N will include

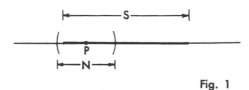

Fig. 1

at least some other point in S. Also we can say that if we define the segment S as "all the points corresponding to all the real numbers greater than 1 and less than 2," we have, as it were, a space—1-dimensional—without its ends. This is in the sense that there is no maximum or minimum (see page 150). This being the case, since we agree that a neighborhood of any point can be as *small* as we choose, we can now say, *Every point in* S *has a neighborhood which is wholly in* S, i.e., that contains no points outside of S. Not that the points in S cannot have neighborhoods *large* enough to contain points outside S, but that if small enough they won't.

Of course we could not apply this reasoning to the end points, but as we know, there aren't any end points in S. It is all a matter of definition; and precise definition keeps cropping up in mathematics. Sometimes lack of definition is basic: the thing is to be precise about whether one is being precise or not.

To recapitulate: as we said on page 150, in a line of points corresponding to the numbers greater than 1 and less than 2, if we choose a point near to 1, we can always choose another between it and 1 (Fig. 2). If someone then puts down another

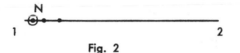

Fig. 2

point still nearer to 1, and claims that it is so near to 1 that there is no more room, we reply that he has attempted the impossible: to put down the *next* point to 1, which cannot be done. Therefore if the point he put down is not next to 1, there is still room for another point, and that will be in the neighborhood—which contains, on both sides of his point, only points in the segment 1–2.

As the reader has probably noticed, this segment is what we referred to as an open set of points. In a closed set: say 1 to 2 inclusive, the above statement about neighborhoods would not apply, because the end points, 1 and 2, would have *no* neighborhoods that did not include at least some point below 1 and above 2, because 1 and 2 would be included in the set.

Limit Points

A limit point—in topology—is not what it sounds like. It does not mean boundary point, although the one can be the other. For example, all the points in a line corresponding to all the real numbers are limit points. A usual (but not the only) definition of a limit point is that *every* neighborhood of it contains another point in the same set of points—in this case the line in question. This sounds strange until we realize that "limit" has a quite special meaning here.

To give a homely, though perhaps uncommon, example: You are standing facing a wall; you step halfway to it, then half the remaining distance, then half that, and so on. If you take 1 second per step you never reach the wall, as there are infinite steps, but nonetheless, the wall is the limit point of your progress. What you are doing, if the total distance is 2 feet, is to step 1 foot, then ½, ¼, ⅛, ¹⁄₁₆ . . . etc., feet; 2 is the total in spite of being the sum of an *infinite series of fractions*. By definition, each time you go only halfway to the wall, never the whole remaining distance, so from this point of view you never reach it. This is the old Achilles-and-the-tortoise wrangle: does he catch the tortoise?

We all know that he does—mauger the definitions—because he does not take those time-consuming steps. He proceeds at a constant velocity and the graph of his motion is as in Fig. 3. The

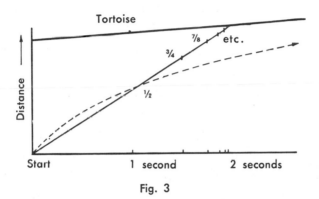

Fig. 3

line of his motion is straight in the graph because his speed is constant, although he passes through the infinite series of fractional positions. If he took 1 second for the first ½ foot, and for each fraction thereafter, the graph would be the curved dotted line, and as far as time is concerned, he would never reach the tortoise—the slanted line at the top. If, however, we add the *times actually taken by Achilles* for the succeeding fractional distances, they come to a finite total: 2 times the time taken for the first half.

The thing that makes the tortoise and the wall limit points is that they can be reached—or approached—in this closer-and-ever-closer way, this continuous process: *not that we can jump it.* We can, in other words, get as close as we choose— which is very arbitrary, but very important in topology. An example of a limit point in 2 dimensions is here given as a problem. No higher mathematics is needed for the proof: only elementary geometry and ordinary logic.

A man is on a flat plain: he walks due west 1 mile, then due north ½ mile, then east ¼ mile, then south ⅛ mile, then west half that, and so on. He keeps making a sharp right turn every time, and every time goes half the previous distance, so that he proceeds in a kind of square spiral, as shown in Fig. 4.

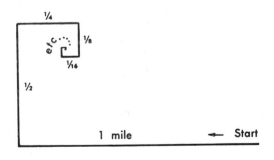

Fig. 4

Where does he end up? It is easy to see he walks altogether 2 miles, so if he keeps a constant speed he gets there: can the exact spot be found—designated—by using ruler and compass? Answer in the Appendix.

Still another sort of limit point—involving time but not space, since position plays no part—would be the moment of being saved just in time if you were on a raft in the sea. You had taken a ship that went down: will the rescuers get to you "in time"? That "time" would be a limit point; but if it had been a question of taking that ship or another, it would not be. It is context that counts: 2 is not a limit point *as one of the integers,* but it is as the sum of 1, ½, ¼, etc., or a member of all the real numbers between say, —1 and +3.

II

Sets

Valid or Merely True?

Bertrand Russell has said that in mathematics we never know what we are talking about. This means that nowadays mathematics is not concerned with truth, but with validity. Not with whether it necessarily applies to the world as it is, but whether it makes logical sense within its own boundaries and according to its own rules.

In the old days it was thought that Euclid's geometry applied to the space around us, but since Einstein the telescopes have shown that space is not Euclidean when we take a big enough sample. Had it turned out otherwise it would not have made Euclid's geometry either less or more valid: merely accidentally more applicable, and mathematics is not concerned with applicability. It is, of course, pleasant when mathematics can be applied.

For this reason one must not misunderstand a

mathematician who seems to talk wildly about infinity: he does not intend to know *what* he is talking about—at least not in terms of any actual infinity, if such there be. Just as when a geometer talks of a right angle he is not suggesting that there is a truly exact one anywhere: he is referring to The Right Angle—an idea, or an ideal, about which certain reasoning can logically be made. In mathematics there are different infinities, and different contexts in which we reason about them. It is the relations that count—not so much the things among which the relationships exist. Topologists finally take leave of their senses, as do most mathematicians: at first topology deals with things perceived by our senses, then more and more with the subject of *how they deal with them.* Finally the sensible objects—the things originally under discussion—are left out entirely. Mathematicians often do their best work this way.

As we saw before, topology aims at the invariant in things; but the things have to be referred to somehow, if very generally, and the best way to refer to things in this way, and yet retain the kind of relationships that exist with topological invariants, is by treating them in groups, or *sets.* We shall now show how we can manipulate sets without having the slightest idea what they are sets of.

Venn Diagrams

Let us start with a set of books, some hard-cover and some paperback, in a shelf. We shall ignore their order and also how many there are, but in this case we do know what we are talking about: books—not all the books in the world, but our set, S. Some are paperbacks, and these form a subset of S, and we shall call this subset P. The way this is usually defined is that *if P is a subset of S, then every element of P is an element of S*—here element means example: in this case, book. To save space it is written $P \subset S$. The hard-covers also form a subset: $H \subset S$. Some of the books are foreign, and they are included in both the subsets P and H. We show a diagrammatic representation of this (Fig. 1), in which the sets and subsets are shown as areas, but the positions and relative amounts of the elements are not indicated. Obvi-

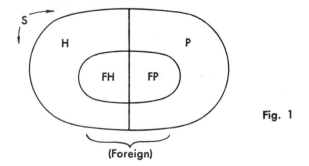

Fig. 1

(Foreign)

ously a book cannot be both paperback and hard-cover, so these two sets—a subset can be considered as a set in its own right—do not overlap. They are mutually exclusive: *disjoint.* Here we see the large area *S* is divided into *H* and *P,* both of these subsets being overlapped by the smaller oval of the foreign books, forming two subsets of *H* and *P: FH* and *FP* in Fig. 1.

We can make another diagram to express who has read what. To do this we shall use the more orthodox *Venn diagram* (Fig. 2): *M*=read by man, and *W*=read by wife. The overlap shows the books both have read, and this area is the *intersection* of *M* and *W:* written *M* ∩ *W.* The unread books are the area of *S* outside *M* and *W.* To describe the set of all the books that have been read we speak of the *sum* (or union) of *M* and *W.* This we write, *M* ∪ *W,* and it includes the intersection *M* ∩ *W.* It means "either-or-both."

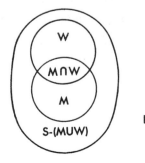

Fig. 2

165

Thus the set of the unread books can be written $S — (M \cup W)$. We express the fact that two sets —or subsets—are disjoint by writing, of Fig. 1: $H \cap P = 0$, meaning that H and P do not intersect.

Venn diagrams can be used for demonstrating certain logical relationships: (1) If all books are printed, and (2) All printing is in ink, then (3) All books are in ink. In Fig. 3 the only books to be

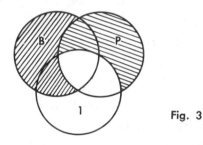

Fig. 3

considered are in the intersection $B \cap P$, as there are no unprinted books (1), so we block out the part of B not in $B \cap P$. Then we block out the part of P not in $P \cap I$, as there is no printing except in ink (2). From what is left we see that all the remaining B is in $B \cap I$; or *all books are in ink.* This is so simple as to be trite—but when we try it with four subsets, *none of which are mutually exclusive* (like read and unread), it gets complicated. It is in fact a minor puzzle merely to draw the Venn diagram for four sets, remembering that every possible combination—one, two, three, or all four at a time—must appear. In doing this it is best to work out first a list of all the combinations: there are 15—always one less than 2 raised to the power of n, where n is the number of sets: $2^n - 1$, unless we include the combination of none of them: then it is 2^n. (It is only fair to mention that the sets need not be shown as even approximate circles: they may have to be long ovals. Answer on page 173.)

A subset of this empty kind above, that has no elements, like a set of unprinted books on imaginary paper, is called a null set. This is regarded as an artificial convention by some topologists, and can lead to complications if always insisted on, but it is useful in algebraic sets. An example of how it has a certain validity that doesn't seem entirely algebraic would be this: in the game of Twenty Questions if someone chooses "the hole in a doughnut," and the question is asked, "Is it vegetable?" (as opposed to mineral), it is hard to answer Yes or No, as it might well be thought to depend on what surrounds the doughnut—air or milk. After all, its very shape is made up *by,* if not of, doughnut material, so perhaps the best answer is that it is a null set of doughnut particles. The game won't allow such an answer—possibly just as well.

Apart from the complexity caused by many intersecting sets, some of the theorems involving only three sets are oddly subtle, and do not spring to mind. Take for example this one:

$$A \cup (B \cap C) = (A \cup B) \cap (A \cup C)$$

This means: "The *sum of A* and the *intersection of B with C* equals the *intersection* of the *sum of A and B* with the *sum of A and C*." Put like that it is a little hard to imagine. In everyday terms it could mean: "The group of all people who are Addlepated, or Blond–Curlyhaired—*or* both—comprises all who are simultaneously in the group of Addlepated or Blond or both, *and* the group of Addlepated or Curlyhaired or both." This has a dubious sound: the first part emphasizes *either-or,* and the second emphasizes *and.* Is English grammar letting us down? Rigorous concentration on the precise meaning of the sentence can induce partial hypnosis, but with a Venn diagram all becomes clear, Fig. 4. In Fig. 5 we have all the circle *A* (all the Addlepated). In Fig. 6 we have the intersection *B* ∩ *C* (Blond–Curlyhaired, not *B or C*). In Fig. 7 we add them: $A \cup (B \cap C)$. Then we follow a like process for the second term of the equation: in Fig. 8 we have all of *A* and *B* (including their intersection—the group comprising the

Addlepated or Blond or both), and in Fig. 9 all of
A and *C* (the group Addlepated or Curlyhaired or
both). In Fig. 10 we superimpose Figs. 8 and 9 to
see what overlaps, and the result is identical with
Fig. 7. *Q.E.D.*

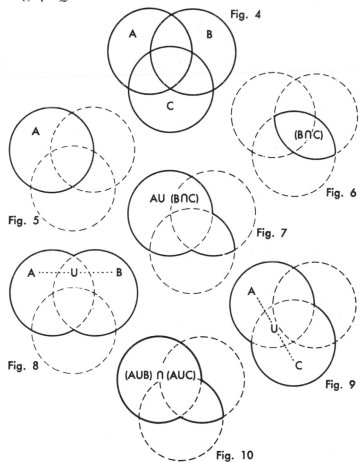

Fig. 4

Fig. 5

Fig. 6

AU (B∩C)

Fig. 7

Fig. 8

Fig. 9

(AUB) ∩ (AUC)

Fig. 10

With Venn diagrams it is sometimes better to use the blocking-out method, as on page 166. From Fig. 1, page 164, we would get Fig. 11, in which

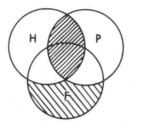

Fig. 11

we block out the intersection of hard-cover and paperback, as no book is both: $H \cap P = 0$. Then we block out all foreign books outside H and P, as we have no unbound books: $F - (H \cup P) = 0$. This does not tell us anything we did not already know, so Fig. 11 is clear enough. Then again, Figs. 4–10 could lead to trouble if the equation had the first term reversed in order; $(B \cap C) \cup A$. We might unfortunately start by blocking out all of the parts of B and C that did not intersect (Fig. 12), and when we add the result to A (*all* of A),

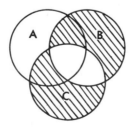

Fig. 12

we have to erase the parts of A just blocked out. Thus using heavy outlines, or at least a progressive series of figures, is sometimes clearer. To recapitulate:

1. A *subset* of a set S is the set all of whose elements are also elements of the set $S: A \subset S$. (When we say that an element p is in, or *belongs* to, S, we write $p \in S$.)
2. The *sum* of two sets A and B is a set all of whose elements are in A or B or both: $A \cup B$.
3. The *intersection* of sets A and B is a set all of whose elements are in both A and $B: A \cap B$.
4. The *complement* of a set A, when A is a subset of S, is that part—or all those elements—of S that are not in $A: S-A$.

The idea of a complement in topology is important: we can at once apply it to a disk: if the disk is a blackened area of a surface, its complement is the unblackened part. The complement of a set S in the universe can be written $U-S$. It becomes still more important when we apply sets to topology. In sets, as in geometry, we find a number of theorems—not necessarily obvious, yet rather unimportant-seeming—which nevertheless help to build a coherent and viable whole. We list a few that the reader can prove with Venn diagrams. (*Note:* It might be a good idea to copy this list and display it prominently until the symbols are familiar: $\subset$ is the *subset* of; $\in$ is an *element* of; $\cup$ is the *sum;* $\cap$ is the *intersection; $A-B$* is the *complement* of B in A.)

1. If $A \subset B$, then $A \cup B = B$.
2. If $A \subset B$, then $A \cap B = A$.
3. If $A \subset B$ and $B \subset C$, then $A \subset C$.
4. $(A \cap B) \cap C = A \cap (B \cap C)$.
5. $A \cap (B \cup C) = (A \cap B) \cup (A \cap C)$.

Before going on, we must give warning that the symbols used here are not universal. If the reader pursues the subject in other books he will find different symbols used for the same things. We list some of them here:

Either or both:

(the shaded parts)

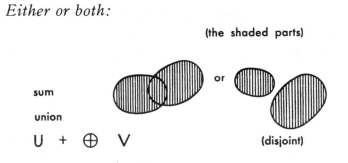

sum

union

U + ⊕ V

or

(disjoint)

Both:

intersection

product

common part

∩ · ⊙ ∧   AB (merely placed together)

Is an element of:

∈ , sometimes ∈ , which is so small
it can be mistaken for 𝜀, or epsilon

Answer to puzzle on pages 166–167 is in Fig. 15. Now try it with five sets. Answer is in the Appendix.

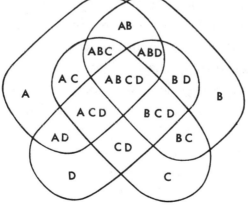

Fig. 15

Open and Closed Sets

So far we have referred to sets of people, or books: in topology sets usually consist of points. When these points mean what they do in geometry, topolgists call the space in which they lie, *Euclidean*. In this case a plane is in itself a 2-dimensional space—frequently written E^2. A line would be E^1, and the others would be E^3, and so on to E^n. The spaces we are now going to discuss are Euclidean, and they will be *metric* spaces. Before explaining this term, we must keep in mind that our goal is always toward the general, and thus we hope to find relationships and theorems that apply to any space, without reference to measurement, and the points will no longer be Euclidean points but *any unspecified things* to which we can apply these relationships meaningfully. This kind of generalization is natural to mathematics: six pairs are a dozen, whether loaves or days.

When topology deals with metric spaces it means that the points have a certain order, in the sense that if the *distance* between point a and point b is zero, then $a=b$. Also, that the sum of the distances between a and b, and a and c is either equal to b-to-c or greater (Fig. 16). (Equal only if $a=b$ and/or $b=c$.) This is pretty simple, but note that it does not specifically say *how much* bigger: only bigger; or the same; or smaller. Nevertheless it does bring in distance, and topology strives as a rule not to. Here it is not breaking its own rules—merely using its own methods with regard to a special kind of case. When these methods are applied to spaces without any distance reference, the question of open and closed becomes paramount. Paramount because the distinction survives all distortion—since distance is left out.

Fig. 16

Let us consider a space consisting of a set of points in a line; then the simplest example of an open set is "all the points corresponding to numbers *greater than* 0 and *less than* 1." A closed set is "all the points corresponding to the numbers *equal to or greater than* 0 and *equal to or less than* 1. But we can get better definitions.

For "open" we repeat more or less what was said on pages 156–157: A point set (set of points) is *open if every point has a neighborhood entirely in the set*. With the comments on page 157 in mind this is not hard to see. If the universe in which our set lies has a single dimension—is an infinite line—then the open set is a segment with its end points missing. (End points are also called boundary points.)

For "closed," the definition is not, however, the simple converse of this. It is: *A set is closed if it contains all its limit points.* It sounds reasonable, but why the sudden switch to limit points? The word to be emphasized in it is *all*. The previous, open set consisted of nothing but limit points, but the missing end points, although not *in* the set (by definition), were limit points *of* the set in that they are approachable within the set, in the way we defined limit points (pages 158–161). A point *in* the set can be as close as we please to either of the end points: therefore, they are limit points of the set. If they are now added to the set—made members of it by definition—the set now contains *all* its limit points, and is closed.

If we had defined this original lack merely in terms of end points (boundary points), it would not allow another fine-spun distinction. If our set-without-end-points were not in a single-dimension universe, but in a *plane,* then clearly none of its points would have any neighborhoods that lay entirely within the set, because neighborhoods would no longer be parts of a line, but 2-dimensional areas (Fig. 17). This is because this space-in-general—this universe—has 2 dimensions, even

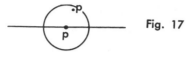

Fig. 17

if our set has not, and these neighborhoods would necessarily contain points like p' that are not in the line at all. Thus the set is *not* open. Is it, then, closed? The answer, rather surprisingly, is No. Closed and open are not paired, mutually exclusive but necessary attributes. Our set is neither, and that is another reason for bringing limit points into the definition of "closed."

The above set was *not* closed, because of failing to include all its limit points—the end points, here—so it is neither open nor closed. Can a set be both closed and open? The answer is Yes: the whole plane in the above example is both. Being infinite it has no end points, so it cannot be said to include them: therefore it *is* open. But by the same token it includes all the points that there are, and therefore its limit points, so it *is also* closed. Most extraordinary.

Thus we have in a plane: (1) *open sets,* e.g., the interior of a given area—say a triangle—because since it does not include its boundary, all its points can have neighborhoods entirely within the set; (2) *closed sets,* e.g., the triangle plus its boundary; (3) *both open and closed,* e.g., the whole infinite space—in any number of chosen dimensions; (4) *neither open nor closed,* e.g., the interior of a 2-dimensional area in a 3-dimensional space, for reasons the same as given for a segment of a line in a plane. (Add to the list of alternative symbols and names: an open set is also a domain.)

Using the idea of Venn diagrams but without drawing anything, we can reason that the complement of an open set is closed: $U - S$ is closed with regard to S (nothing is said about its outer, nonexisting boundaries). It means that since the boundary points are not in S, they must be included in $U - S$. The reverse of this is that the complement of a closed set is open, for similar reasons. The reader can prove to his own satisfaction that the sum of two open sets is open, and that the intersection of two closed sets is closed.

With regard to an open set, we can infer the following: the sum of it and its limit points is called its *closure* (written $\overline{S}$); so $\overline{S}$ is closed, because it now contains all the limit points.

Neighborhoods, being defined the way they are, can be considered as open—though some topologists say they are closed, too, on grounds too abstruse to go into here. The former sounds somehow more hospitable.

We have seen sets consisting entirely of limit points: how about sets without them, or at least not only of limit points? There is one that we mentioned before, on page 158, although we did not speak of it as a set: namely the points corresponding to 1, ½, ¼, ⅛ . . . etc., on the Euclidean line. We could just as well take 1, ½, ⅓, ¼ . . . and this, too, would be a set, none of whose points would be a limit point *except the last one,* 0. We can say this because with the exception of 0 every point can have a neighborhood small enough so as not to contain any other of these points. The end point, 0, is a limit point because it is the point to which the series converges, and we can approach it in this way.

We can go further and say that if the set consists of these fractional points that converge to 0, then *none of the points* corresponding to all the real numbers on the Euclidean line are limit points *of this set* (except 0). This is so according to the reasoning we used about the "next" point, page 150: if we cannot name or truly specify a point next to one of the member points of our set, we cannot name one which cannot have a neighborhood small enough so that it fails to contain a member of the set. Complicated, but inescapable.

The set just mentioned had only one limit point:

can we find one without any? The answer is Yes: the set of points corresponding to, say, 1, ½, ⅓ . . . to ⅒, in the Euclidean line, has no limit points, for fairly obvious reasons. The inevitable presence of a limit point—or at least one, and not necessarily *in* the set—when the set is *infinite* is proved by a neat theorem that can be intuitively grasped. The Bolzano-Weierstrass theorem says that we can represent an infinite set as an area in which we know there is an infinite number of points . . . *somewhere* in it. They may or may not be spread over it evenly (for example, in the last case the infinite clumping was at the zero end only). We draw a line arbitrarily through the middle (Fig. 18), and there must be infinitely many on one side if not both. Let us say it is on the right. We then divide that side with another line (Fig. 19), and ask the same question: now it's on top. We go on subdividing (Fig. 20): every time getting closer, and at last closing in on the point

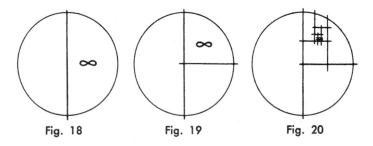

Fig. 18   Fig. 19   Fig. 20

where infinity is lurking. This is by definition a limit point.

If the limit point had been (again by definition) outside the set, we would just as surely have found it. It would of course be in the boundary or closure of the set, as in our previous example had we decided to leave zero out of it. On the other hand, if we had an infinite set like the Euclidean line—say all the real numbers from 1 to 2—which has infinite points in any neighborhood, then we would have infinite points on both sides of the lines as we put them down, and we could have chosen either side at every stage, and thus find a limit point anywhere throughout the entire set—again finding that it consists entirely of limit points.

The fact that there is no one to tell us which side we must take at each stage is immaterial: there *is* such a series of correct placements, and this proves the presence of the limit point.

Another noteworthy attribute of an infinite set is that it can have a subset which can be made to correspond one-for-one to *all of the set*. Contradictory as this may seem, we shall consider an infinite set of books, numbered 1, 2, 3, etc., and we shall take the even-numbered books as our subset: $E \subset B$. We now enumerate E: the first, second, third, and so on, even-numbered books. We now pair these with the numbers of the whole set of books, B: first goes with 1, second with 2, third with 3, and so on ad infinitum. The fact that there would seem to be more books than there are even-numbered books is neither here nor there: we will never run out of the latter to pair off with the former. The set is endless, and so is the subset.

Transformations

A transformation is a relatedness between two things. It is also called a function. The way it is used in a sentence is confusing: sometimes as a noun (as above) and sometimes as an adjective. It may apply to the process of transforming, or the rule governing it, or the result, or the thing transformed *and* the result. The last would be called a function rather than a transformation. Earlier, in speaking of invariants, we said they survived distortion: in this context, distortion is a transformation.

But the word has a more general sense: when we paired even-numbered books with all the books, that was a transformation, too. Here the word "function" would refer to what is called *ordered pairs*—in this case, books, but they might be points. Whenever one says that every element of a set is going to be changed into, or made to *correspond to,* the elements of another set, that is a transformation. We can go back to the torus that was molded (transformed) into a teacup-with-handle and label every point on the torus and draw, as it were, an arrow from each to its corresponding point on the teacup. In this case we might identify the points-in-general around the hole in the teacup handle as having originated somewhere around the hole in the torus, but beyond that it would be hard to tell, particularly when it came to which part had become, say, the rim of the cup.

This correspondence is at once too hard and fast, and too vague. That is to say we must know when, and when not, to be specific. For example, when we say Fig. 21 is equivalent to Fig. 22, there are only two identifiable points of one in terms of the other; p and p'. It will be noticed that the *image*—the corresponding point in the second set—is referred to as $f(p)$, or $f(p')$, meaning *function p:* here used as an adjective. We cannot say that any other point chosen in Fig. 21 corresponds to any particular one in Fig. 22, although we *can* say that *no* point on the closed curve of Fig. 21 corresponds to *any* point on the semidetached line of Fig. 22. However, in this transformation certain things

Fig. 21   Fig. 22

do remain: the point where the line joins the closed curve, and the point where the line ends, or has a free vertex, and they can be so identified.

Earlier, when Mr. Jones and Dr. Situs were reminding one another that the figures of Fig. 23

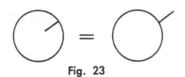

Fig. 23

are identical, they had this rather generalized identification in mind. It harks back to our saying (page 5) that a permissible distortion allows cutting, if things are rejoined.

For example we can see how Figs. 24–25 are topologically equivalent. They both consist of two closed curves, *a* and *b,* and *a* has a vertex *c* where it joins a line *d,* which has another vertex *e* where it joins *b*. From *c* runs another line *f* with a vertex *g* at its free end. What other identifications can be made? (Answer at end of chapter.)

These rather different-sounding kinds of transformation are alike by analogy, but there is more

184

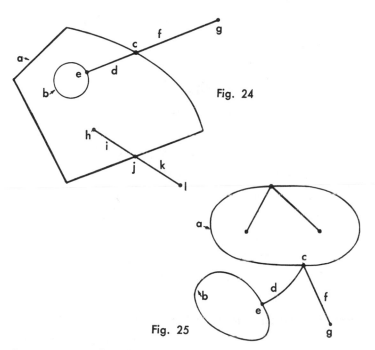

Fig. 24

Fig. 25

in common. In either case we make a correspondence between points according to some agreed-upon rule or law, or function—*f*. If there is a function, or transformation of set *A* into set *B* (Fig. 26), we say that the point *p* in *A* has an

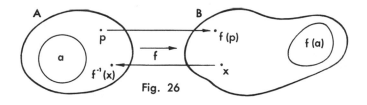

Fig. 26

image, $f(p)$, in B. This also applies to subsets: $a \subset A$ has an image, $f(a) \subset B$. It is also used in reverse: f^{-1} is the symbol, and $f^{-1}(x)$ means here the point in A of which x is the image, and is also called the inverse image of x.

We said earlier that the word "continuous" was usually confined to processes (page 152), and that means functions. A *continuous transformation* means that points sufficiently close together in A will have points close together in B: this is to be taken in the same spirit as our remarks about neighborhoods on pages 154–155. As before, we can improve on this definition. In a continuous transformation of set A into set B, any subset a of A that is *open* has an image, $f(a)$, in B that is *also open*. Examination of the definition of "open" (page 175) will show the implications of this: a set (or subset) is open if *every point* has a neighborhood entirely in the set. Without going into fearsome detail, we can see that this guarantees the possibility of an infinite degree of closeness (as in the case of a limit point) of a pair of points in A *and* their images in B. The main thing is that continuity leads to *mapping* (not to be confused with the maps of Chapter 6, though they are examples of mappings).

A PRIMER

SEE, the LADY has FAL-LEN in the WA-TER!

Can she SWIM?

No, but she can FLOAT—Here she comes NOW!

What are those NUM-BERS?

They are the var-ious POINTS on her HEAD—

Why has she got POINTS on her HEAD?

They are IM-AG-IN-ARY: They show which parts of her Head made the Pretty RIP-PLES. See, the RIP-PLES are NUM-BERED.

Why do the Num-bers go BACK-WARD?

Because that is the ORDER they were made in. FIRST the TOP of her HEAD made the OUT-SIDE Rip-ple—That's NUM-BER ONE.

How did it get to be on the OUT-SIDE?

It Started FIRST. Then the circles on her HEAD made the SMAL-LER RIP-PLES, count-ing IN.

I don't see any CIR-CLES on her head.

They are Imag-in-ary, too. As her HEAD comes UP it makes—

What is the FISH doing?

NOTH-ING.

Mapping

A map can be fantastically unlike its inverse image, even when quite Euclidean and geometrical in the everyday sense. As can be imagined from the sample page just given, if we map all the concentric circles drawn on a hemisphere by projecting them onto a flat surface, all sorts of different results can be got, depending on where we put the point of origin of the projection lines.

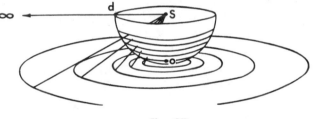

Fig. 27

If we drew the projection lines from *S* in Fig. 27, and the hemisphere were sitting on the plane as shown, the circles would be projected onto the plane in the same ever-widening way they are on the hemisphere, except that as we got nearer to the diameter they would be extending out much further in proportion to their inverse images, and at *d* itself they would be at infinity. The ones at the bottom would be nearer to scale, and the bottom point, *o,* would be touching its counterpart. Fig 28,

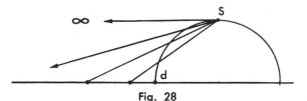

Fig. 28

on the other hand, shows what happens when the hemisphere is inverted: the diameter coincides with its image, and when we go up, where the circles on the hemisphere get smaller, the images get larger, and the top point has an image that is an infinite circle at infinity. This projection gives results somewhat similar to the Lady in the Water, where the top point became a circle also, but the increasing radii were in closer proportion to the inverse images.

All of the foregoing are mappings: any coherent rule that takes the elements of one set and puts them into another is a mapping, even when there is no question of geometrical projection, provided that the process is continuous, *one-to-one* and ditto in reverse. *One-to-one* means that every point of set A becomes one and only one point in set B. This can work in reverse, so that the transformation backward, f^{-1}, takes each point in set B to one and only one point in A. This symmetrical and continuous function is our old friend homeomorphism, now properly defined.

A graph is a kind of transformation, too. It is not necessary to be familiar with analytic geometry to know what a graph is: a temperature chart, for example, is a graph that identifies moments in time with degrees Fahrenheit. It takes the elements—or a chosen few—from set T (time), and relates them to the corresponding elements in set F (degrees Fahrenheit). In actuality the nurse doesn't take one's temperature continuously, but it could be done with a recording thermometer which made a line on paper that would be a continuous graph. Again we can define "graph" better.

Here the idea of proximity, or neighborhood, is a little different from the more or less circular shapes we were using in E^2, (2-dimensional space), although this involves 2 dimensions also. We think in terms of the single dimension of set T, and then of the single dimension of set F—separate criteria which are compared, for this kind of graph is a *comparison* of two sets, each of 1 dimension. Then we can say that a graph is continuous if it is continuous at every point. It will be continuous at a given point p (Fig. 29) when *any* pair of horizontal lines is drawn, one above and one below p, if we can draw a vertical pair, one to the right and one to the left of p, *that does not contain any points not contained by the horizontals*. This is

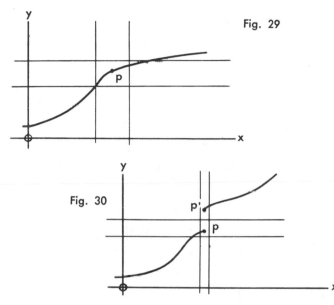

Fig. 29

Fig. 30

plainer if we show a counterexample (Fig. 30), which shows a discontinuous graph—it has a gap in it, but that gap has no meaning in terms of x . . . (there is no gap in the *set x*)—and if we drew the horizontals shown, no matter how close together we put the verticals on either side of p, p' will be included because it is vertically above p, and p' is *not* between the horizontals. It might be thought that a vertical graph line would contradict this, but such lines are not considered fair game in a topological function, as they cause confusion, and spoil the definition—a rather sneaky way to avoid trouble, but it leads to usefulness and clarity.

The above is like the definition of a continuous transformation on page 186: but here the neighborhood of *p,* when it is considered *as an element of set y* (*Ny*), is a space along the *y*-axis—an up-and-down qualification, while its neighborhood considered *as an element of set x* is *Nx,* a horizontal one. The latter can, if we want, be labeled $f(x)$, if this f is known to mean this kind of graphing.

A graph, when it is continuous and one-to-one both ways, is not so much a map as a mapping: it shows not just set *x* and its map set *y,* but the relationship of *x* and *y,* in—so to say—graphic form. This sort of graph has 2 dimensions, because each of the sets it relates has 1 dimension. If we made a graph of the changes in a moving picture, it would have 3 dimensions. One can see that some of the more complicated transformations would lead to completely undrawable graphs of *n* dimensions, such as graphing the points in—not on—a sphere into the points in a cube—6 dimensions.

Homotopy

"Homeomorphism," now better defined, has a kind of functional attendant in the word "homotopy." This specifies not only the possibility of a homeomorphic transformation, but the circumstances in which it is possible. Any closed curve, we have kept saying, is deformable into any other, but even though we allow ourselves a certain leeway in the transformation of the first of these into the second, i.e., cutting and rejoining (Fig. 31),

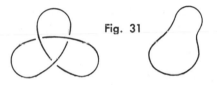

Fig. 31

it must be understood that there may be contexts which preclude this. As soon as we specify the type of space *in which the deformation is to happen,* we have imposed a new condition. If the closed curve is string, and we have open space to maneuver in, we can do things to it that would not be possible were the curve confined to a plane by law.

It is true that by one set of rules we change *A* to *B* in a plane (Fig. 32), but only because the plane is not mentioned as the theater of operations—the plane was vaguely and erroneously assumed: the figures are *planar;* that's all. In a different context

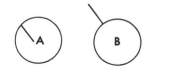

Fig. 32

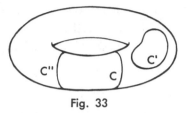

Fig. 33

(Fig. 33), it is meaningful to say that the closed curve C on the torus cannot be transformed to C' *on the torus,* although it can be to C''. Out in E^3 there would be no trouble whatsoever—if the curve were wire, or imaginary.

The Jordan curve theorem, mentioned before, would apply on the surface of the torus in the case of C', but obviously not of C'' or C, as they do not divide the surface. The criterion in this case is that on a torus C and C'' cannot be continuously contracted to a point, whereas C' can be. This is a property more of the space in which these figures lie than of the figures. We can now say that on a torus any two curves like C and C'' are homotopic, but neither would be to C'.

The transformation itself, particularly if it is a mapping of C into C''—the shaded part in Fig. 34

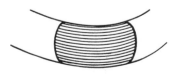

Fig. 34

—is the *homotopy* of C and C″. The idea is carried still further: all intermediate stages are mutually homotopic. In these two somewhat simpler cases of Fig. 35, we can see that a given homotopy is not unique: there are in fact infinitely many that achieve the same results—changing *A* to *B*.

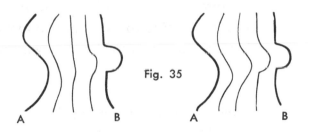

Fig. 35

The mere choice of a particular mapping is less fundamental than the general possibility of there being a mapping or not. None exists for C to C′: an infinity exist for C to C″. All of this may seem to run counter to the anarchy we expected at first as to what is and what is not allowed in topology. But, as was said before, the main thing is to know when to be precise and when to be permissive. Though the confinement implied in homotopy is in a way arbitrary, it allows a further distinction among spaces—a new kind of invariant.

Another criterion we can briefly mention is *compactness*. It refers to a sort of completed infinity: *the* Euclidean plane is not compact, but the surface of a sphere is, though it has as many points. But if we want to define compactness in terms of the set of these points without reference to the surrounding space we use this definition: "A set is compact if every one of its *infinite* subsets has a limit point in the set itself."

This is not quite as obscure as it may at first appear. The reader may care to explain it to himself by using what has been said in the last two chapters.

Answer to the question on page 184: Only one: *j*. It is unequivocally defined as the apex where the lines *i* and *k* meet *a*. But *i* and *k* and their vertices *h* and *l* can *not* be paired with their counterparts, as there is no way of saying which goes with which, except negatively, e.g., *k* is *not h*.

In Conclusion

It frequently happens that when getting a cup of coffee one forgets the cream. The trick, here, is not to go and get the cream, but to take the cup to it. The first way involves four trips: going for the cream, bringing it to the table, taking it back right away, and returning to the coffee. The other way involves two: taking the cup to the refrigerator and returning with the cup. This cannot be helpfully expressed geometrically, but the kind of sequential planning used, though arithmetical, belongs rather in topology.

In the use of electronic computers it is called Programing, and set-theoretic topology is its basis. In designing the fearsomely complicated circuits of these computers they make use of topological network analysis, mentioned in Chapter 8. Topology has found a place in astronomy, also, and indeed many other endeavors where mathematics is needed. These subjects are hardly everyday, but we use topological method in many everyday acts, though unconsciously. Most descriptions of where something is are topological, rather than geometrical: The coat is in your closet; The school is the fourth house beyond the intersection of this street and Route 32; The Pen of my Aunt is in the Garden.

Mariners use geometry, and so do builders, but in ordinary circumstances its use is avoided, except for the metrics of driving distances. and cookery.

The Renaissance marked several changes in scientific thinking and method, one of which is best exemplified by chemistry. Medieval alchemy concentrated on difference in kind—difference in degree seemed less of the essence to them, and their chemistry never got off the ground. With the new ways of thinking, chemistry turned away from the qualitative to the quantitative—from Kind to Degree—and they began to get order from chaos. Mathematics, on the other hand, had always leaned toward the quantitative method—until topology, and the process seems to be reversed.

But not really: in looking back over these pages we can see that while form and measurement are temporarily abandoned they crop up again in a more sophisticated guise, for quality has a quantity; kind has a degree, even if it is not measured with a yardstick. As Stephen Vincent Benét said (he was talking of Lincoln), yardsticks are good for measuring—if you have yards to measure.

Appendix

Page 48. The Gardner solution consists in making an odd number of longitudinal folds in the strip —along its "length," at right angles to the edges we join. Fig. 1 shows the arrangement. If the number of folds is odd, the ends can be brought together with a half-twist to match, no matter how

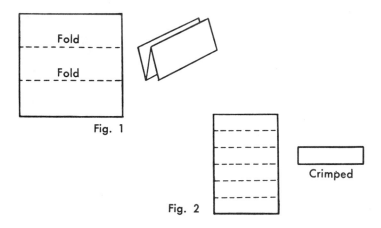

Fig. 1

Fig. 2

Crimped

great the odd number is. This is because an *N* turned upside down is still an *N*, whereas an *M* is altered. Thus an indefinitely wide strip—provided the paper is thin enough—can be crimped into a sufficiently narrow one for us to treat as a conveniently proportioned Moebius strip, Fig. 2. After all our work in Chapter 3 it's infuriating.

Page 49. Before joining the strip, mark it off as in Fig. 3—which is shortened. The numbered points are at the ½ and ¼ positions, and the line is ruled on the back where dotted. When joined, the line will be straight and continuous. The cut starts at 1 and proceeds numerically: after 5, when x is reached the strip will be found to have opened out, so the cut will have to continue on the line. The cut thus seems not to be continuous, but the line *was*.

When 8 is reached the strip will come apart

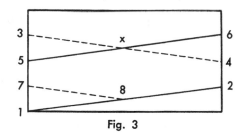

Fig. 3

into two pieces of equal area, which is proved by counting the triangles in Fig. 4. The shaded part is one piece and the unshaded part the other; each contains 8. Figures 15 and 16 on page 49 show a method that also divides the strip into two pieces of equal area, but the cut does not start at the edge. The cut is at the ¼ mark from the edge.

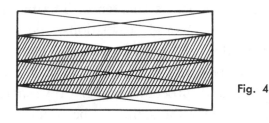

Fig. 4

Page 72. The pattern is shown in Fig. 5 and is best laid out on heavy square-ruled paper, with about an inch and a half to a unit. The solid lines are cut, and the dotted lines folded. The frequent half units are due to this pattern being made symmetrical, except for unit number 14 and its opposite number, *x*. The perspective views identify the numbered and lettered units, thus giving a guide for the folding after the cuts are made. Joints are butt, not overlap, so cellulose tape is used.

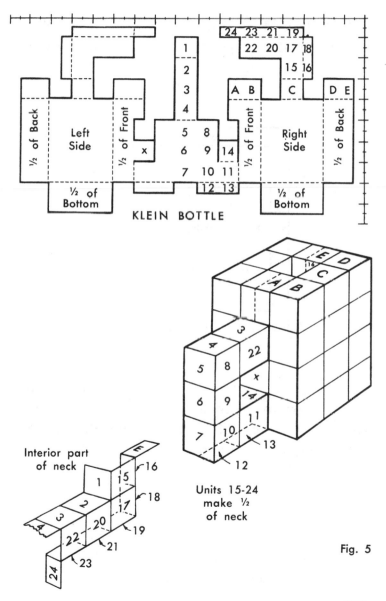

½ of Back

Left Side

½ of Front

x

½ of Front

Right Side

½ of Back

½ of Bottom

½ of Bottom

KLEIN BOTTLE

24 23 21 19
22 20 17 18
15 16

A B   C   D E

1
2
3
4
5 8
6 9 14
7 10 11
12 13

E D
C
A B

3
4 22
5 8 x
6 9 14
7 10 11
12 13

Interior part of neck

E
1 15 16
2 17 18
3 20 19
4 22 21
24 23

Units 15-24 make ½ of neck

Fig. 5

203

Page 119. Mix ⅓ of the red with all the blue, getting enough purple for 16 square feet. Fig. 6 shows the color plan.

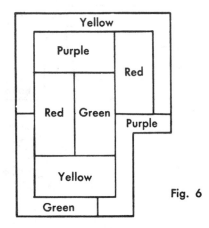

Fig. 6

Page 131. The series *is* a combination of two, but it boils down to one formula: Where n is the number of partial cuts before the final cut, the number of pieces is $2^n + 1$: (3, 5, 9, 17, etc.).

Page 161. Construction (Fig. 7).

Draw the lines 1–2, 2–3, and 3–4, making 2 and 3 right angles. Join 1 to 3, and 2 to 4: the intersection C is the required limit point.

(Each succeeding
line of the man's
route is half of
the previous one.)

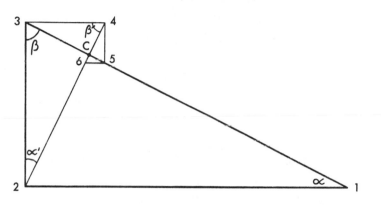

Fig. 7

Proof.

Add lines 4–5 and 5–6, making 4 and 5 right angles.

Line 1–2 is twice line 2–3, and 2–3 is twice 3–4;
∴ triangles 1–2–3 and 2–3–4 are similar,
∴ $\angle \alpha = \angle \alpha'$, and $\angle \beta = \angle \beta'$.
∴ Triangle 2–3–C is similar to triangle 1–2–3;
∴ the angles at C are right angles.

All succeeding right-angle triangles in the following list are similar: 1–2–C, 2–3–C, 3–4–C, 4–5–C . . . etc., and their hypotenuses, since they are successively halved, form the required route taken by the man. *Q.E.D.*

Appendix

Page 173. Fig. 8 is a diagram for five sets: dotted lines show beginning of the sixth, carried around the border of the fifth. An unlimited number of sets can be shown this way, each time following the border of the last one.

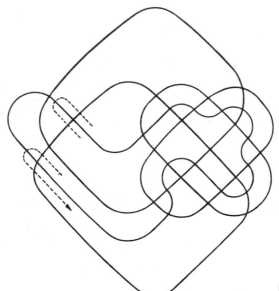

Fig. 8

Index

Index

Index

A CATALOG OF SELECTED
DOVER BOOKS
IN ALL FIELDS OF INTEREST

A CATALOG OF SELECTED
DOVER BOOKS
IN ALL FIELDS OF INTEREST

DRAWINGS OF REMBRANDT, edited by Seymour Slive. Updated Lippmann, Hofstede de Groot edition, with definitive scholarly apparatus. All portraits, biblical sketches, landscapes, nudes. Oriental figures, classical studies, together with selection of work by followers. 550 illustrations. Total of 630pp. 9⅛ × 12¼.
21485-0, 21486-9 Pa., Two-vol. set $29.90

GHOST AND HORROR STORIES OF AMBROSE BIERCE, Ambrose Bierce. 24 tales vividly imagined, strangely prophetic, and decades ahead of their time in technical skill: "The Damned Thing," "An Inhabitant of Carcosa," "The Eyes of the Panther," "Moxon's Master," and 20 more. 199pp. 5⅜ × 8½. 20767-6 Pa. $4.95

ETHICAL WRITINGS OF MAIMONIDES, Maimonides. Most significant ethical works of great medieval sage, newly translated for utmost precision, readability. Laws Concerning Character Traits, Eight Chapters, more. 192pp. 5⅜ × 8½.
24522-5 Pa. $5.95

THE EXPLORATION OF THE COLORADO RIVER AND ITS CANYONS, J. W. Powell. Full text of Powell's 1,000-mile expedition down the fabled Colorado in 1869. Superb account of terrain, geology, vegetation, Indians, famine, mutiny, treacherous rapids, mighty canyons, during exploration of last unknown part of continental U.S. 400pp. 5⅜ × 8½. 20094-9 Pa. $7.95

HISTORY OF PHILOSOPHY, Julián Marías. Clearest one-volume history on the market. Every major philosopher and dozens of others, to Existentialism and later. 505pp. 5⅜ × 8½. 21739-6 Pa. $9.95

ALL ABOUT LIGHTNING, Martin A. Uman. Highly readable nontechnical survey of nature and causes of lightning, thunderstorms, ball lightning, St. Elmo's Fire, much more. Illustrated. 192pp. 5⅜ × 8½. 25237-X Pa. $5.95

SAILING ALONE AROUND THE WORLD, Captain Joshua Slocum. First man to sail around the world, alone, in small boat. One of great feats of seamanship told in delightful manner. 67 illustrations. 294pp. 5⅜ × 8½. 20326-3 Pa. $4.95

LETTERS AND NOTES ON THE MANNERS, CUSTOMS AND CONDITIONS OF THE NORTH AMERICAN INDIANS, George Catlin. Classic account of life among Plains Indians: ceremonies, hunt, warfare, etc. 312 plates. 572pp. of text. 6⅛ × 9¼. 22118-0, 22119-9, Pa., Two-vol. set $17.90

THE SECRET LIFE OF SALVADOR DALÍ, Salvador Dalí. Outrageous but fascinating autobiography through Dalí's thirties with scores of drawings and sketches and 80 photographs. A must for lovers of 20th-century art. 432pp. 6½ × 9¼. (Available in U.S. only) 27454-3 Pa. $9.95

CATALOG OF DOVER BOOKS

THE BOOK OF BEASTS: Being a Translation from a Latin Bestiary of the Twelfth Century, T. H. White. Wonderful catalog of real and fanciful beasts: manticore, griffin, phoenix, amphivius, jaculus, many more. White's witty erudite commentary on scientific, historical aspects enhances fascinating glimpse of medieval mind. Illustrated. 296pp. 5⅜ × 8¼. (Available in U.S. only) 24609-4 Pa. $7.95

FRANK LLOYD WRIGHT: Architecture and Nature with 160 Illustrations, Donald Hoffmann. Profusely illustrated study of influence of nature—especially prairie—on Wright's designs for Fallingwater, Robie House, Guggenheim Museum, other masterpieces. 96pp. 9¼ × 10¾. 25098-9 Pa. $8.95

FRANK LLOYD WRIGHT'S FALLINGWATER, Donald Hoffmann. Wright's famous waterfall house: planning and construction of organic idea. History of site, owners, Wright's personal involvement. Photographs of various stages of building. Preface by Edgar Kaufmann, Jr. 100 illustrations. 112pp. 9¼ × 10.
23671-4 Pa. $8.95

YEARS WITH FRANK LLOYD WRIGHT: Apprentice to Genius, Edgar Tafel. Insightful memoir by a former apprentice presents a revealing portrait of Wright the man, the inspired teacher, the greatest American architect. 372 black-and-white illustrations. Preface. Index. vi + 228pp. 8¼ × 11. 24801-1 Pa. $10.95

THE STORY OF KING ARTHUR AND HIS KNIGHTS, Howard Pyle. Enchanting version of King Arthur fable has delighted generations with imaginative narratives of exciting adventures and unforgettable illustrations by the author. 41 illustrations. xviii + 313pp. 6⅛ × 9¼. 21445-1 Pa. $6.95

THE GODS OF THE EGYPTIANS, E. A. Wallis Budge. Thorough coverage of numerous gods of ancient Egypt by foremost Egyptologist. Information on evolution of cults, rites and gods; the cult of Osiris; the Book of the Dead and its rites; the sacred animals and birds; Heaven and Hell; and more. 956pp. 6⅛ × 9¼.
22055-9, 22056-7 Pa., Two-vol. set $21.90

A THEOLOGICO-POLITICAL TREATISE, Benedict Spinoza. Also contains unfinished *Political Treatise*. Great classic on religious liberty, theory of government on common consent. R. Elwes translation. Total of 421pp. 5⅜ × 8½.
20249-6 Pa. $7.95

INCIDENTS OF TRAVEL IN CENTRAL AMERICA, CHIAPAS, AND YUCATAN, John L. Stephens. Almost single-handed discovery of Maya culture; exploration of ruined cities, monuments, temples; customs of Indians. 115 drawings. 892pp. 5⅜ × 8½. 22404-X, 22405-8 Pa., Two-vol. set $17.90

LOS CAPRICHOS, Francisco Goya. 80 plates of wild, grotesque monsters and caricatures. Prado manuscript included. 183pp. 6⅛ × 9⅜. 22384-1 Pa. $6.95

AUTOBIOGRAPHY: The Story of My Experiments with Truth, Mohandas K. Gandhi. Not hagiography, but Gandhi in his own words. Boyhood, legal studies, purification, the growth of the Satyagraha (nonviolent protest) movement. Critical, inspiring work of the man who freed India. 480pp. 5⅜ × 8½. (Available in U.S. only)
24593-4 Pa. $6.95

ILLUSTRATED DICTIONARY OF HISTORIC ARCHITECTURE, edited by Cyril M. Harris. Extraordinary compendium of clear, concise definitions for over 5,000 important architectural terms complemented by over 2,000 line drawings. Covers full spectrum of architecture from ancient ruins to 20th-century Modernism. Preface. 592pp. 7½ × 9⅝. 24444-X Pa. $15.95

THE NIGHT BEFORE CHRISTMAS, Clement Moore. Full text, and woodcuts from original 1848 book. Also critical, historical material. 19 illustrations. 40pp. 4⅝ × 6. 22797-9 Pa. $2.50

THE LESSON OF JAPANESE ARCHITECTURE: 165 Photographs, Jiro Harada. Memorable gallery of 165 photographs taken in the 1930's of exquisite Japanese homes of the well-to-do and historic buildings. 13 line diagrams. 192pp. 8⅜ × 11¼. 24778-3 Pa. $10.95

THE AUTOBIOGRAPHY OF CHARLES DARWIN AND SELECTED LETTERS, edited by Francis Darwin. The fascinating life of eccentric genius composed of an intimate memoir by Darwin (intended for his children); commentary by his son, Francis; hundreds of fragments from notebooks, journals, papers; and letters to and from Lyell, Hooker, Huxley, Wallace and Henslow. xi + 365pp. 5⅜ × 8. 20479-0 Pa. $6.95

WONDERS OF THE SKY: Observing Rainbows, Comets, Eclipses, the Stars and Other Phenomena, Fred Schaaf. Charming, easy-to-read poetic guide to all manner of celestial events visible to the naked eye. Mock suns, glories, Belt of Venus, more. Illustrated. 299pp. 5¼ × 8¼. 24402-4 Pa. $7.95

BURNHAM'S CELESTIAL HANDBOOK, Robert Burnham, Jr. Thorough guide to the stars beyond our solar system. Exhaustive treatment. Alphabetical by constellation: Andromeda to Cetus in Vol. 1; Chamaeleon to Orion in Vol. 2; and Pavo to Vulpecula in Vol. 3. Hundreds of illustrations. Index in Vol. 3. 2,000pp. 6⅛ × 9¼. 23567-X, 23568-8, 23673-0 Pa., Three-vol. set $41.85

STAR NAMES: Their Lore and Meaning, Richard Hinckley Allen. Fascinating history of names various cultures have given to constellations and literary and folkloristic uses that have been made of stars. Indexes to subjects. Arabic and Greek names. Biblical references. Bibliography. 563pp. 5⅜ × 8½. 21079-0 Pa. $8.95

THIRTY YEARS THAT SHOOK PHYSICS: The Story of Quantum Theory, George Gamow. Lucid, accessible introduction to influential theory of energy and matter. Careful explanations of Dirac's anti-particles, Bohr's model of the atom, much more. 12 plates. Numerous drawings. 240pp. 5⅜ × 8½. 24895-X Pa. $5.95

CHINESE DOMESTIC FURNITURE IN PHOTOGRAPHS AND MEASURED DRAWINGS, Gustav Ecke. A rare volume, now affordably priced for antique collectors, furniture buffs and art historians. Detailed review of styles ranging from early Shang to late Ming. Unabridged republication. 161 black-and-white drawings, photos. Total of 224pp. 8⅜ × 11¼. (Available in U.S. only) 25171-3 Pa. $13.95

VINCENT VAN GOGH: A Biography, Julius Meier-Graefe. Dynamic, penetrating study of artist's life, relationship with brother, Theo, painting techniques, travels, more. Readable, engrossing. 160pp. 5⅜ × 8½. (Available in U.S. only) 25253-1 Pa. $4.95

HOW TO WRITE, Gertrude Stein. Gertrude Stein claimed anyone could understand her unconventional writing—here are clues to help. Fascinating improvisations, language experiments, explanations illuminate Stein's craft and the art of writing. Total of 414pp. 4⅝ × 6⅜. 23144-5 Pa. $6.95

ADVENTURES AT SEA IN THE GREAT AGE OF SAIL: Five Firsthand Narratives, edited by Elliot Snow. Rare true accounts of exploration, whaling, shipwreck, fierce natives, trade, shipboard life, more. 33 illustrations. Introduction. 353pp. 5⅜ × 8½. 25177-2 Pa. $8.95

THE HERBAL OR GENERAL HISTORY OF PLANTS, John Gerard. Classic descriptions of about 2,850 plants—with over 2,700 illustrations—includes Latin and English names, physical descriptions, varieties, time and place of growth, more. 2,706 illustrations. xlv + 1,678pp. 8½ × 12¼. 23147-X Cloth. $75.00

DOROTHY AND THE WIZARD IN OZ, L. Frank Baum. Dorothy and the Wizard visit the center of the Earth, where people are vegetables, glass houses grow and Oz characters reappear. Classic sequel to *Wizard of Oz.* 256pp. 5⅜ × 8. 24714-7 Pa. $5.95

SONGS OF EXPERIENCE: Facsimile Reproduction with 26 Plates in Full Color, William Blake. This facsimile of Blake's original "Illuminated Book" reproduces 26 full-color plates from a rare 1826 edition. Includes "The Tyger," "London," "Holy Thursday," and other immortal poems. 26 color plates. Printed text of poems. 48pp. 5¼ × 7. 24636-1 Pa. $3.95

SONGS OF INNOCENCE, William Blake. The first and most popular of Blake's famous "Illuminated Books," in a facsimile edition reproducing all 31 brightly colored plates. Additional printed text of each poem. 64pp. 5¼ × 7. 22764-2 Pa. $3.95

PRECIOUS STONES, Max Bauer. Classic, thorough study of diamonds, rubies, emeralds, garnets, etc.: physical character, occurrence, properties, use, similar topics. 20 plates, 8 in color. 94 figures. 659pp. 6⅛ × 9¼. 21910-0, 21911-9 Pa., Two-vol. set $15.90

ENCYCLOPEDIA OF VICTORIAN NEEDLEWORK, S. F. A. Caulfeild and Blanche Saward. Full, precise descriptions of stitches, techniques for dozens of needlecrafts—most exhaustive reference of its kind. Over 800 figures. Total of 679pp. 8⅛ × 11. Two volumes. Vol. 1 22800-2 Pa. $11.95 Vol. 2 22801-0 Pa. $11.95

THE MARVELOUS LAND OF OZ, L. Frank Baum. Second Oz book, the Scarecrow and Tin Woodman are back with hero named Tip, Oz magic. 136 illustrations. 287pp. 5⅜ × 8½. 20692-0 Pa. $5.95

WILD FOWL DECOYS, Joel Barber. Basic book on the subject, by foremost authority and collector. Reveals history of decoy making and rigging, place in American culture, different kinds of decoys, how to make them, and how to use them. 140 plates. 156pp. 7⅞ × 10¾. 20011-6 Pa. $8.95

HISTORY OF LACE, Mrs. Bury Palliser. Definitive, profusely illustrated chronicle of lace from earliest times to late 19th century. Laces of Italy, Greece, England, France, Belgium, etc. Landmark of needlework scholarship. 266 illustrations. 672pp. 6⅛ × 9¼. 24742-2 Pa. $14.95

CATALOG OF DOVER BOOKS

ILLUSTRATED GUIDE TO SHAKER FURNITURE, Robert Meader. All furniture and appurtenances, with much on unknown local styles. 235 photos. 146pp. 9 × 12. 22819-3 Pa. $8.95

WHALE SHIPS AND WHALING: A Pictorial Survey, George Francis Dow. Over 200 vintage engravings, drawings, photographs of barks, brigs, cutters, other vessels. Also harpoons, lances, whaling guns, many other artifacts. Comprehensive text by foremost authority. 207 black-and-white illustrations. 288pp. 6 × 9. 24808-9 Pa. $9.95

THE BERTRAMS, Anthony Trollope. Powerful portrayal of blind self-will and thwarted ambition includes one of Trollope's most heartrending love stories. 497pp. 5⅜ × 8½. 25119-5 Pa. $9.95

ADVENTURES WITH A HAND LENS, Richard Headstrom. Clearly written guide to observing and studying flowers and grasses, fish scales, moth and insect wings, egg cases, buds, feathers, seeds, leaf scars, moss, molds, ferns, common crystals, etc.—all with an ordinary, inexpensive magnifying glass. 209 exact line drawings aid in your discoveries. 220pp. 5⅜ × 8½. 23330-8 Pa. $4.95

RODIN ON ART AND ARTISTS, Auguste Rodin. Great sculptor's candid, wide-ranging comments on meaning of art; great artists; relation of sculpture to poetry, painting, music; philosophy of life, more. 76 superb black-and-white illustrations of Rodin's sculpture, drawings and prints. 119pp. 8⅜ × 11¼. 24487-3 Pa. $7.95

FIFTY CLASSIC FRENCH FILMS, 1912–1982: A Pictorial Record, Anthony Slide. Memorable stills from Grand Illusion, Beauty and the Beast, Hiroshima, Mon Amour, many more. Credits, plot synopses, reviews, etc. 160pp. 8¼ × 11. 25256-6 Pa. $11.95

THE PRINCIPLES OF PSYCHOLOGY, William James. Famous long course complete, unabridged. Stream of thought, time perception, memory, experimental methods; great work decades ahead of its time. 94 figures. 1,391pp. 5⅜ × 8½. 20381-6, 20382-4 Pa., Two-vol. set $23.90

BODIES IN A BOOKSHOP, R. T. Campbell. Challenging mystery of blackmail and murder with ingenious plot and superbly drawn characters. In the best tradition of British suspense fiction. 192pp. 5⅜ × 8½. 24720-1 Pa. $4.95

CALLAS: PORTRAIT OF A PRIMA DONNA, George Jellinek. Renowned commentator on the musical scene chronicles incredible career and life of the most controversial, fascinating, influential operatic personality of our time. 64 black-and-white photographs. 416pp. 5⅜ × 8¼. 25047-4 Pa. $8.95

GEOMETRY, RELATIVITY AND THE FOURTH DIMENSION, Rudolph Rucker. Exposition of fourth dimension, concepts of relativity as Flatland characters continue adventures. Popular, easily followed yet accurate, profound. 141 illustrations. 133pp. 5⅜ × 8½. 23400-2 Pa. $4.95

HOUSEHOLD STORIES BY THE BROTHERS GRIMM, with pictures by Walter Crane. 53 classic stories—Rumpelstiltskin, Rapunzel, Hansel and Gretel, the Fisherman and his Wife, Snow White, Tom Thumb, Sleeping Beauty, Cinderella, and so much more—lavishly illustrated with original 19th century drawings. 114 illustrations. x + 269pp. 5⅜ × 8½. 21080-4 Pa. $4.95

SUNDIALS, Albert Waugh. Far and away the best, most thorough coverage of ideas, mathematics concerned, types, construction, adjusting anywhere. Over 100 illustrations. 230pp. 5⅜ × 8½. 22947-5 Pa. $5.95

PICTURE HISTORY OF THE NORMANDIE: With 190 Illustrations, Frank O. Braynard. Full story of legendary French ocean liner: Art Deco interiors, design innovations, furnishings, celebrities, maiden voyage, tragic fire, much more. Extensive text. 144pp. 8⅞ × 11¼. 25257-4 Pa. $10.95

THE FIRST AMERICAN COOKBOOK: A Facsimile of "American Cookery," 1796, Amelia Simmons. Facsimile of the first American-written cookbook published in the United States contains authentic recipes for colonial favorites—pumpkin pudding, winter squash pudding, spruce beer, Indian slapjacks, and more. Introductory Essay and Glossary of colonial cooking terms. 80pp. 5⅜ × 8½. 24710-4 Pa. $3.50

101 PUZZLES IN THOUGHT AND LOGIC, C. R. Wylie, Jr. Solve murders and robberies, find out which fishermen are liars, how a blind man could possibly identify a color—purely by your own reasoning! 107pp. 5⅜ × 8½. 20367-0 Pa. $2.95

ANCIENT EGYPTIAN MYTHS AND LEGENDS, Lewis Spence. Examines animism, totemism, fetishism, creation myths, deities, alchemy, art and magic, other topics. Over 50 illustrations. 432pp. 5⅜ × 8½. 26525-0 Pa. $8.95

ANTHROPOLOGY AND MODERN LIFE, Franz Boas. Great anthropologist's classic treatise on race and culture. Introduction by Ruth Bunzel. Only inexpensive paperback edition. 255pp. 5⅜ × 8½. 25245-0 Pa. $7.95

THE TALE OF PETER RABBIT, Beatrix Potter. The inimitable Peter's terrifying adventure in Mr. McGregor's garden, with all 27 wonderful, full-color Potter illustrations. 55pp. 4¼ × 5½. (Available in U.S. only) 22827-4 Pa. $1.75

THREE PROPHETIC SCIENCE FICTION NOVELS, H. G. Wells. *When the Sleeper Wakes, A Story of the Days to Come* and *The Time Machine* (full version). 335pp. 5⅜ × 8½. (Available in U.S. only) 20605-X Pa. $8.95

APICIUS COOKERY AND DINING IN IMPERIAL ROME, edited and translated by Joseph Dommers Vehling. Oldest known cookbook in existence offers readers a clear picture of what foods Romans ate, how they prepared them, etc. 49 illustrations. 301pp. 6⅛ × 9¼. 23563-7 Pa. $7.95

SHAKESPEARE LEXICON AND QUOTATION DICTIONARY, Alexander Schmidt. Full definitions, locations, shades of meaning of every word in plays and poems. More than 50,000 exact quotations. 1,485pp. 6½ × 9¼. 22726-X, 22727-8 Pa., Two-vol. set $31.90

THE WORLD'S GREAT SPEECHES, edited by Lewis Copeland and Lawrence W. Lamm. Vast collection of 278 speeches from Greeks to 1970. Powerful and effective models; unique look at history. 842pp. 5⅜ × 8½. 20468-5 Pa. $12.95

THE BLUE FAIRY BOOK, Andrew Lang. The first, most famous collection, with many familiar tales: Little Red Riding Hood, Aladdin and the Wonderful Lamp, Puss in Boots, Sleeping Beauty, Hansel and Gretel, Rumpelstiltskin; 37 in all. 138 illustrations. 390pp. 5⅜ × 8½. 21437-0 Pa. $6.95

THE STORY OF THE CHAMPIONS OF THE ROUND TABLE, Howard Pyle. Sir Launcelot, Sir Tristram and Sir Percival in spirited adventures of love and triumph retold in Pyle's inimitable style. 50 drawings, 31 full-page. xviii + 329pp. 6½ × 9¼. 21883-X Pa. $7.95

THE MYTHS OF THE NORTH AMERICAN INDIANS, Lewis Spence. Myths and legends of the Algonquins, Iroquois, Pawnees and Sioux with comprehensive historical and ethnological commentary. 36 illustrations. 5⅜ × 8½. 25967-6 Pa. $8.95

GREAT DINOSAUR HUNTERS AND THEIR DISCOVERIES, Edwin H. Colbert. Fascinating, lavishly illustrated chronicle of dinosaur research, 1820s to 1960. Achievements of Cope, Marsh, Brown, Buckland, Mantell, Huxley, many others. 384pp. 5¼ × 8¼. 24701-5 Pa. $7.95

THE TASTEMAKERS, Russell Lynes. Informal, illustrated social history of American taste 1850s–1950s. First popularized categories Highbrow, Lowbrow, Middlebrow. 129 illustrations. New (1979) afterword. 384pp. 6 × 9. 23993-4 Pa. $8.95

DOUBLE CROSS PURPOSES, Ronald A. Knox. A treasure hunt in the Scottish Highlands, an old map, unidentified corpse, surprise discoveries keep reader guessing in this cleverly intricate tale of financial skullduggery. 2 black-and-white maps. 320pp. 5⅜ × 8½. (Available in U.S. only) 25032-6 Pa. $6.95

AUTHENTIC VICTORIAN DECORATION AND ORNAMENTATION IN FULL COLOR: 46 Plates from "Studies in Design," Christopher Dresser. Superb full-color lithographs reproduced from rare original portfolio of a major Victorian designer. 48pp. 9¼ × 12¼. 25083-0 Pa. $7.95

PRIMITIVE ART, Franz Boas. Remains the best text ever prepared on subject, thoroughly discussing Indian, African, Asian, Australian, and, especially, Northern American primitive art. Over 950 illustrations show ceramics, masks, totem poles, weapons, textiles, paintings, much more. 376pp. 5⅜ × 8. 20025-6 Pa. $7.95

SIDELIGHTS ON RELATIVITY, Albert Einstein. Unabridged republication of two lectures delivered by the great physicist in 1920–21. *Ether and Relativity* and *Geometry and Experience*. Elegant ideas in nonmathematical form, accessible to intelligent layman. vi + 56pp. 5⅜ × 8½. 24511-X Pa. $3.95

THE WIT AND HUMOR OF OSCAR WILDE, edited by Alvin Redman. More than 1,000 ripostes, paradoxes, wisecracks: Work is the curse of the drinking classes, I can resist everything except temptation, etc. 258pp. 5⅜ × 8½. 20602-5 Pa. $4.95

ADVENTURES WITH A MICROSCOPE, Richard Headstrom. 59 adventures with clothing fibers, protozoa, ferns and lichens, roots and leaves, much more. 142 illustrations. 232pp. 5⅜ × 8½. 23471-1 Pa. $4.95

PLANTS OF THE BIBLE, Harold N. Moldenke and Alma L. Moldenke. Standard reference to all 230 plants mentioned in Scriptures. Latin name, biblical reference, uses, modern identity, much more. Unsurpassed encyclopedic resource for scholars, botanists, nature lovers, students of Bible. Bibliography. Indexes. 123 black-and-white illustrations. 384pp. 6 × 9. 25069-5 Pa. $8.95

FAMOUS AMERICAN WOMEN: A Biographical Dictionary from Colonial Times to the Present, Robert McHenry, ed. From Pocahontas to Rosa Parks, 1,035 distinguished American women documented in separate biographical entries. Accurate, up-to-date data, numerous categories, spans 400 years. Indices. 493pp. 6½ × 9¼. 24523-3 Pa. $10.95

THE FABULOUS INTERIORS OF THE GREAT OCEAN LINERS IN HISTORIC PHOTOGRAPHS, William H. Miller, Jr. Some 200 superb photographs capture exquisite interiors of world's great "floating palaces"—1890s to 1980s: *Titanic, Ile de France, Queen Elizabeth, United States, Europa,* more. Approx. 200 black-and-white photographs. Captions. Text. Introduction. 160pp. 8⅞ × 11¾. 24756-2 Pa. $9.95

THE GREAT LUXURY LINERS, 1927–1954: A Photographic Record, William H. Miller, Jr. Nostalgic tribute to heyday of ocean liners. 186 photos of *Ile de France, Normandie, Leviathan, Queen Elizabeth, United States,* many others. Interior and exterior views. Introduction. Captions. 160pp. 9 × 12. 24056-8 Pa. $12.95

A NATURAL HISTORY OF THE DUCKS, John Charles Phillips. Great landmark of ornithology offers complete detailed coverage of nearly 200 species and subspecies of ducks: gadwall, sheldrake, merganser, pintail, many more. 74 full-color plates, 102 black-and-white. Bibliography. Total of 1,920pp. 8⅝ × 11¼. 25141-1, 25142-X Cloth., Two-vol. set $100.00

THE SEAWEED HANDBOOK: An Illustrated Guide to Seaweeds from North Carolina to Canada, Thomas F. Lee. Concise reference covers 78 species. Scientific and common names, habitat, distribution, more. Finding keys for easy identification. 224pp. 5⅜ × 8½. 25215-9 Pa. $6.95

THE TEN BOOKS OF ARCHITECTURE: The 1755 Leoni Edition, Leon Battista Alberti. Rare classic helped introduce the glories of ancient architecture to the Renaissance. 68 black-and-white plates. 336pp. 8⅜ × 11¼. 25239-6 Pa. $14.95

MISS MACKENZIE, Anthony Trollope. Minor masterpieces by Victorian master unmasks many truths about life in 19th-century England. First inexpensive edition in years. 392pp. 5⅜ × 8½. 25201-9 Pa. $8.95

THE RIME OF THE ANCIENT MARINER, Gustave Doré, Samuel Taylor Coleridge. Dramatic engravings considered by many to be his greatest work. The terrifying space of the open sea, the storms and whirlpools of an unknown ocean, the ice of Antarctica, more—all rendered in a powerful, chilling manner. Full text. 38 plates. 77pp. 9¼ × 12. 22305-1 Pa. $4.95

THE EXPEDITIONS OF ZEBULON MONTGOMERY PIKE, Zebulon Montgomery Pike. Fascinating firsthand accounts (1805–6) of exploration of Mississippi River, Indian wars, capture by Spanish dragoons, much more. 1,088pp. 5⅜ × 8½. 25254-X, 25255-8 Pa., Two-vol. set $25.90

A CONCISE HISTORY OF PHOTOGRAPHY: Third Revised Edition, Helmut Gernsheim. Best one-volume history—camera obscura, photochemistry, daguerreotypes, evolution of cameras, film, more. Also artistic aspects—landscape, portraits, fine art, etc. 281 black-and-white photographs. 26 in color. 176pp. 8⅜ × 11¼.
25128-4 Pa. $14.95

THE DORÉ BIBLE ILLUSTRATIONS, Gustave Doré. 241 detailed plates from the Bible: the Creation scenes, Adam and Eve, Flood, Babylon, battle sequences, life of Jesus, etc. Each plate is accompanied by the verses from the King James version of the Bible. 241pp. 9 × 12. 23004-X Pa. $9.95

WANDERINGS IN WEST AFRICA, Richard F. Burton. Great Victorian scholar/adventurer's invaluable descriptions of African tribal rituals, fetishism, culture, art, much more. Fascinating 19th-century account. 624pp. 5⅜ × 8½. 26890-X Pa. $12.95

HISTORIC HOMES OF THE AMERICAN PRESIDENTS, Second Revised Edition, Irvin Haas. Guide to homes occupied by every president from Washington to Bush. Visiting hours, travel routes, more. 175 photos. 160pp. 8¼ × 11.
26751-2 Pa. $9.95

THE HISTORY OF THE LEWIS AND CLARK EXPEDITION, Meriwether Lewis and William Clark, edited by Elliott Coues. Classic edition of Lewis and Clark's day-by-day journals that later became the basis for U.S. claims to Oregon and the West. Accurate and invaluable geographical, botanical, biological, meteorological and anthropological material. Total of 1,508pp. 5⅜ × 8½.
21268-8, 21269-6, 21270-X Pa., Three-vol. set $29.85

LANGUAGE, TRUTH AND LOGIC, Alfred J. Ayer. Famous, clear introduction to Vienna, Cambridge schools of Logical Positivism. Role of philosophy, elimination of metaphysics, nature of analysis, etc. 160pp. 5⅜ × 8½. (Available in U.S. and Canada only) 20010-8 Pa. $3.95

MATHEMATICS FOR THE NONMATHEMATICIAN, Morris Kline. Detailed, college-level treatment of mathematics in cultural and historical context, with numerous exercises. For liberal arts students. Preface. Recommended Reading Lists. Tables. Index. Numerous black-and-white figures. xvi + 641pp. 5⅜ × 8½.
24823-2 Pa. $11.95

HANDBOOK OF PICTORIAL SYMBOLS, Rudolph Modley. 3,250 signs and symbols, many systems in full; official or heavy commercial use. Arranged by subject. Most in Pictorial Archive series. 143pp. 8⅛ × 11. 23357-X Pa. $7.95

INCIDENTS OF TRAVEL IN YUCATAN, John L. Stephens. Classic (1843) exploration of jungles of Yucatan, looking for evidences of Maya civilization. Travel adventures, Mexican and Indian culture, etc. Total of 669pp. 5⅜ × 8½.
20926-1, 20927-X Pa., Two-vol. set $13.90

CATALOG OF DOVER BOOKS

DEGAS: An Intimate Portrait, Ambroise Vollard. Charming, anecdotal memoir by famous art dealer of one of the greatest 19th-century French painters. 14 black-and-white illustrations. Introduction by Harold L. Van Doren. 96pp. 5⅜ × 8½.
25131-4 Pa. $4.95

PERSONAL NARRATIVE OF A PILGRIMAGE TO AL-MADINAH AND MECCAH, Richard F. Burton. Great travel classic by remarkably colorful personality. Burton, disguised as a Moroccan, visited sacred shrines of Islam, narrowly escaping death. 47 illustrations. 959pp. 5⅜ × 8½.
21217-3, 21218-1 Pa., Two-vol. set $19.90

PHRASE AND WORD ORIGINS, A. H. Holt. Entertaining, reliable, modern study of more than 1,200 colorful words, phrases, origins and histories. Much unexpected information. 254pp. 5⅜ × 8½.
20758-7 Pa. $5.95

THE RED THUMB MARK, R. Austin Freeman. In this first Dr. Thorndyke case, the great scientific detective draws fascinating conclusions from the nature of a single fingerprint. Exciting story, authentic science. 320pp. 5⅜ × 8½. (Available in U.S. only)
25210-8 Pa. $6.95

AN EGYPTIAN HIEROGLYPHIC DICTIONARY, E. A. Wallis Budge. Monumental work containing about 25,000 words or terms that occur in texts ranging from 3000 B.C. to 600 A.D. Each entry consists of a transliteration of the word, the word in hieroglyphs, and the meaning in English. 1,314pp. 6⅜ × 10.
23615-3, 23616-1 Pa., Two-vol. set $35.90

THE COMPLEAT STRATEGYST: Being a Primer on the Theory of Games of Strategy, J. D. Williams. Highly entertaining classic describes, with many illustrated examples, how to select best strategies in conflict situations. Prefaces. Appendices. xvi + 268pp. 5⅜ × 8½.
25101-2 Pa. $6.95

THE ROAD TO OZ, L. Frank Baum. Dorothy meets the Shaggy Man, little Button-Bright and the Rainbow's beautiful daughter in this delightful trip to the magical Land of Oz. 272pp. 5⅜ × 8.
25208-6 Pa. $5.95

POINT AND LINE TO PLANE, Wassily Kandinsky. Seminal exposition of role of point, line, other elements in nonobjective painting. Essential to understanding 20th-century art. 127 illustrations. 192pp. 6½ × 9¼.
23808-3 Pa. $5.95

LADY ANNA, Anthony Trollope. Moving chronicle of Countess Lovel's bitter struggle to win for herself and daughter Anna their rightful rank and fortune—perhaps at cost of sanity itself. 384pp. 5⅜ × 8½.
24669-8 Pa. $8.95

EGYPTIAN MAGIC, E. A. Wallis Budge. Sums up all that is known about magic in Ancient Egypt: the role of magic in controlling the gods, powerful amulets that warded off evil spirits, scarabs of immortality, use of wax images, formulas and spells, the secret name, much more. 253pp. 5⅜ × 8½.
22681-6 Pa. $4.95

THE DANCE OF SIVA, Ananda Coomaraswamy. Preeminent authority unfolds the vast metaphysic of India: the revelation of her art, conception of the universe, social organization, etc. 27 reproductions of art masterpieces. 192pp. 5⅜ × 8½.
24817-8 Pa. $6.95

CHRISTMAS CUSTOMS AND TRADITIONS, Clement A. Miles. Origin, evolution, significance of religious, secular practices. Caroling, gifts, yule logs, much more. Full, scholarly yet fascinating; non-sectarian. 400pp. 5⅜ × 8½.
23354-5 Pa. $7.95

THE HUMAN FIGURE IN MOTION, Eadweard Muybridge. More than 4,500 stopped-action photos, in action series, showing undraped men, women, children jumping, lying down, throwing, sitting, wrestling, carrying, etc. 390pp. 7⅞ × 10⅝.
20204-6 Cloth. $24.95

THE MAN WHO WAS THURSDAY, Gilbert Keith Chesterton. Witty, fast-paced novel about a club of anarchists in turn-of-the-century London. Brilliant social, religious, philosophical speculations. 128pp. 5⅜ × 8½. 25121-7 Pa. $3.95

A CÉZANNE SKETCHBOOK: Figures, Portraits, Landscapes and Still Lifes, Paul Cézanne. Great artist experiments with tonal effects, light, mass, other qualities in over 100 drawings. A revealing view of developing master painter, precursor of Cubism. 102 black-and-white illustrations. 144pp. 8¾ × 6⅜. 24790-2 Pa. $6.95

AN ENCYCLOPEDIA OF BATTLES: Accounts of Over 1,560 Battles from 1479 B.C. to the Present, David Eggenberger. Presents essential details of every major battle in recorded history, from the first battle of Megiddo in 1479 B.C. to Grenada in 1984. List of Battle Maps. New Appendix covering the years 1967–1984. Index. 99 illustrations. 544pp. 6½ × 9¼. 24913-1 Pa. $14.95

AN ETYMOLOGICAL DICTIONARY OF MODERN ENGLISH, Ernest Weekley. Richest, fullest work, by foremost British lexicographer. Detailed word histories. Inexhaustible. Total of 856pp. 6½ × 9¼.
21873-2, 21874-0 Pa., Two-vol. set $19.90

WEBSTER'S AMERICAN MILITARY BIOGRAPHIES, edited by Robert McHenry. Over 1,000 figures who shaped 3 centuries of American military history. Detailed biographies of Nathan Hale, Douglas MacArthur, Mary Hallaren, others. Chronologies of engagements, more. Introduction. Addenda. 1,033 entries in alphabetical order. xi + 548pp. 6½ × 9¼. (Available in U.S. only)
24758-9 Pa. $13.95

LIFE IN ANCIENT EGYPT, Adolf Erman. Detailed older account, with much not in more recent books: domestic life, religion, magic, medicine, commerce, and whatever else needed for complete picture. Many illustrations. 597pp. 5⅜ × 8½.
22632-8 Pa. $9.95

HISTORIC COSTUME IN PICTURES, Braun & Schneider. Over 1,450 costumed figures shown, covering a wide variety of peoples: kings, emperors, nobles, priests, servants, soldiers, scholars, townsfolk, peasants, merchants, courtiers, cavaliers, and more. 256pp. 8⅜ × 11¼. 23150-X Pa. $9.95

THE NOTEBOOKS OF LEONARDO DA VINCI, edited by J. P. Richter. Extracts from manuscripts reveal great genius; on painting, sculpture, anatomy, sciences, geography, etc. Both Italian and English. 186 ms. pages reproduced, plus 500 additional drawings, including studies for *Last Supper, Sforza* monument, etc. 860pp. 7⅞ × 10¾. (Available in U.S. only) 22572-0, 22573-9 Pa., Two-vol. set $35.90

THE ART NOUVEAU STYLE BOOK OF ALPHONSE MUCHA: All 72 Plates from "Documents Decoratifs" in Original Color, Alphonse Mucha. Rare copyright-free design portfolio by high priest of Art Nouveau. Jewelry, wallpaper, stained glass, furniture, figure studies, plant and animal motifs, etc. Only complete one-volume edition. 80pp. 9⅜ × 12¼. 24044-4 Pa. $9.95

ANIMALS: 1,419 COPYRIGHT-FREE ILLUSTRATIONS OF MAMMALS, BIRDS, FISH, INSECTS, ETC., edited by Jim Harter. Clear wood engravings present, in extremely lifelike poses, over 1,000 species of animals. One of the most extensive pictorial sourcebooks of its kind. Captions. Index. 284pp. 9 × 12. 23766-4 Pa. $9.95

OBELISTS FLY HIGH, C. Daly King. Masterpiece of American detective fiction, long out of print, involves murder on a 1935 transcontinental flight—"a very thrilling story"—NY Times. Unabridged and unaltered republication of the edition published by William Collins Sons & Co. Ltd., London, 1935. 288pp. 5⅜ × 8½. (Available in U.S. only) 25036-9 Pa. $5.95

VICTORIAN AND EDWARDIAN FASHION: A Photographic Survey, Alison Gernsheim. First fashion history completely illustrated by contemporary photographs. Full text plus 235 photos, 1840–1914, in which many celebrities appear. 240pp. 6½ × 9¼. 24205-6 Pa. $8.95

THE ART OF THE FRENCH ILLUSTRATED BOOK, 1700–1914, Gordon N. Ray. Over 630 superb book illustrations by Fragonard, Delacroix, Daumier, Doré, Grandville, Manet, Mucha, Steinlen, Toulouse-Lautrec and many others. Preface. Introduction. 633 halftones. Indices of artists, authors & titles, binders and provenances. Appendices. Bibliography. 608pp. 8⅜ × 11¼. 25086-5 Pa. $24.95

THE WONDERFUL WIZARD OF OZ, L. Frank Baum. Facsimile in full color of America's finest children's classic. 143 illustrations by W. W. Denslow. 267pp. 5⅜ × 8½. 20691-2 Pa. $7.95

FOLLOWING THE EQUATOR: A Journey Around the World, Mark Twain. Great writer's 1897 account of circumnavigating the globe by steamship. Ironic humor, keen observations, vivid and fascinating descriptions of exotic places. 197 illustrations. 720pp. 5⅜ × 8½. 26113-1 Pa. $15.95

THE FRIENDLY STARS, Martha Evans Martin & Donald Howard Menzel. Classic text marshalls the stars together in an engaging, non-technical survey, presenting them as sources of beauty in night sky. 23 illustrations. Foreword. 2 star charts. Index. 147pp. 5⅜ × 8½. 21099-5 Pa. $3.95

FADS AND FALLACIES IN THE NAME OF SCIENCE, Martin Gardner. Fair, witty appraisal of cranks, quacks, and quackeries of science and pseudoscience: hollow earth, Velikovsky, orgone energy, Dianetics, flying saucers, Bridey Murphy, food and medical fads, etc. Revised, expanded In the Name of Science. "A very able and even-tempered presentation."—The New Yorker. 363pp. 5⅜ × 8. 20394-8 Pa. $6.95

ANCIENT EGYPT: ITS CULTURE AND HISTORY, J. E Manchip White. From pre-dynastics through Ptolemies: society, history, political structure, religion, daily life, literature, cultural heritage. 48 plates. 217pp. 5⅜ × 8½. 22548-8 Pa. $5.95

SIR HARRY HOTSPUR OF HUMBLETHWAITE, Anthony Trollope. Incisive, unconventional psychological study of a conflict between a wealthy baronet, his idealistic daughter, and their scapegrace cousin. The 1870 novel in its first inexpensive edition in years. 250pp. 5⅜ × 8½. 24953-0 Pa. $6.95

LASERS AND HOLOGRAPHY, Winston E. Kock. Sound introduction to burgeoning field, expanded (1981) for second edition. Wave patterns, coherence, lasers, diffraction, zone plates, properties of holograms, recent advances. 84 illustrations. 160pp. 5⅜ × 8¼. (Except in United Kingdom) 24041-X Pa. $3.95

INTRODUCTION TO ARTIFICIAL INTELLIGENCE: Second, Enlarged Edition, Philip C. Jackson, Jr. Comprehensive survey of artificial intelligence—the study of how machines (computers) can be made to act intelligently. Includes introductory and advanced material. Extensive notes updating the main text. 132 black-and-white illustrations. 512pp. 5⅜ × 8½. 24864-X Pa. $10.95

HISTORY OF INDIAN AND INDONESIAN ART, Ananda K. Coomaraswamy. Over 400 illustrations illuminate classic study of Indian art from earliest Harappa finds to early 20th century. Provides philosophical, religious and social insights. 304pp. 6⅜ × 9⅜. 25005-9 Pa. $11.95

THE GOLEM, Gustav Meyrink. Most famous supernatural novel in modern European literature, set in Ghetto of Old Prague around 1890. Compelling story of mystical experiences, strange transformations, profound terror. 13 black-and-white illustrations. 224pp. 5⅜ × 8½. (Available in U.S. only) 25025-3 Pa. $6.95

PICTORIAL ENCYCLOPEDIA OF HISTORIC ARCHITECTURAL PLANS, DETAILS AND ELEMENTS: With 1,880 Line Drawings of Arches, Domes, Doorways, Facades, Gables, Windows, etc., John Theodore Haneman. Sourcebook of inspiration for architects, designers, others. Bibliography. Captions. 141pp. 9 × 12. 24605-1 Pa. $8.95

BENCHLEY LOST AND FOUND, Robert Benchley. Finest humor from early 30s, about pet peeves, child psychologists, post office and others. Mostly unavailable elsewhere. 73 illustrations by Peter Arno and others. 183pp. 5⅜ × 8½. 22410-4 Pa. $4.95

ERTÉ GRAPHICS, Erté. Collection of striking color graphics: *Seasons, Alphabet, Numerals, Aces* and *Precious Stones.* 50 plates, including 4 on covers. 48pp. 9⅜ × 12¼. 23580-7 Pa. $7.95

THE JOURNAL OF HENRY D. THOREAU, edited by Bradford Torrey, F. H. Allen. Complete reprinting of 14 volumes, 1837–61, over two million words; the sourcebooks for *Walden,* etc. Definitive. All original sketches, plus 75 photographs. 1,804pp. 8½ × 12¼. 20312-3, 20313-1 Cloth., Two-vol. set $130.00

CASTLES: Their Construction and History, Sidney Toy. Traces castle development from ancient roots. Nearly 200 photographs and drawings illustrate moats, keeps, baileys, many other features. Caernarvon, Dover Castles, Hadrian's Wall, Tower of London, dozens more. 256pp. 5⅜ × 8¼. 24898-4 Pa. $7.95

AMERICAN CLIPPER SHIPS: 1833–1858, Octavius T. Howe & Frederick C. Matthews. Fully-illustrated, encyclopedic review of 352 clipper ships from the period of America's greatest maritime supremacy. Introduction. 109 halftones. 5 black-and-white line illustrations. Index. Total of 928pp. 5⅜ × 8½.
25115-2, 25116-0 Pa., Two-vol. set $17.90

TOWARDS A NEW ARCHITECTURE, Le Corbusier. Pioneering manifesto by great architect, near legendary founder of "International School." Technical and aesthetic theories, views on industry, economics, relation of form to function, "mass-production spirit," much more. Profusely illustrated. Unabridged translation of 13th French edition. Introduction by Frederick Etchells. 320pp. 6⅛ × 9¼. (Available in U.S. only)
25023-7 Pa. $8.95

THE BOOK OF KELLS, edited by Blanche Cirker. Inexpensive collection of 32 full-color, full-page plates from the greatest illuminated manuscript of the Middle Ages, painstakingly reproduced from rare facsimile edition. Publisher's Note. Captions. 32pp. 9⅜ × 12¼. (Available in U.S. only)
24345-1 Pa. $5.95

BEST SCIENCE FICTION STORIES OF H. G. WELLS, H. G. Wells. Full novel *The Invisible Man*, plus 17 short stories: "The Crystal Egg," "Aepyornis Island," "The Strange Orchid," etc. 303pp. 5⅜ × 8½. (Available in U.S. only)
21531-8 Pa. $6.95

AMERICAN SAILING SHIPS: Their Plans and History, Charles G. Davis. Photos, construction details of schooners, frigates, clippers, other sailcraft of 18th to early 20th centuries—plus entertaining discourse on design, rigging, nautical lore, much more. 137 black-and-white illustrations. 240pp. 6⅛ × 9¼.
24658-2 Pa. $6.95

ENTERTAINING MATHEMATICAL PUZZLES, Martin Gardner. Selection of author's favorite conundrums involving arithmetic, money, speed, etc., with lively commentary. Complete solutions. 112pp. 5⅜ × 8½.
25211-6 Pa. $3.50

THE WILL TO BELIEVE, HUMAN IMMORTALITY, William James. Two books bound together. Effect of irrational on logical, and arguments for human immortality. 402pp. 5⅜ × 8½.
20291-7 Pa. $8.95

THE HAUNTED MONASTERY and THE CHINESE MAZE MURDERS, Robert Van Gulik. 2 full novels by Van Gulik continue adventures of Judge Dee and his companions. An evil Taoist monastery, seemingly supernatural events; overgrown topiary maze that hides strange crimes. Set in 7th-century China. 27 illustrations. 328pp. 5⅜ × 8½.
23502-5 Pa. $6.95

CELEBRATED CASES OF JUDGE DEE (DEE GOONG AN), translated by Robert Van Gulik. Authentic 18th-century Chinese detective novel; Dee and associates solve three interlocked cases. Led to Van Gulik's own stories with same characters. Extensive introduction. 9 illustrations. 237pp. 5⅜ × 8½.
23337-5 Pa. $5.95

Prices subject to change without notice.

Available at your book dealer or write for free catalog to Dept. GI, Dover Publications, Inc., 31 East 2nd St., Mineola, N.Y. 11501. Dover publishes more than 175 books each year on science, elementary and advanced mathematics, biology, music, art, literary history, social sciences and other areas.